Cartography: Visualization of Spatial Data

Cartography:
Visualization
of Spatial Data

MENNO-JAN KRAAK & FERJAN ORMELING

LONGMAN

Addison Wesley Longman Limited
Edinburgh Gate, Harlow
Essex CM20 2JE, England
and Associated Companies throughout the world

First published 1996

British Library Cataloguing in Publication Data
A catalogue entry for this title is available from the British Library.

ISBN 0–582–25953–3

Library of Congress Cataloging-in-Publication Data
A catalog entry for this title is available from the Library of Congress.

Typeset by 16 in 9/11pt Times New Roman

Produced by Longman Singapore Publishers (Pte) Ltd.
Printed in Great Britain by Henry Ling Ltd., at the Dorset Press, Dorchester, Dorset.

Contents

Preface

Our society is one in which we are used to free access to printed maps. General access to maps started with the invention of the printing press and ever since, especially after the demise of the Cold War, more and more cartographic information in printed form has become available. This, in turn, has helped in improving those maps produced: wide circulation entails opening up one's products to criticism and thus to improvement. A book like Jacques Bertin's *Semiology of Graphics* (1967), from which the authors took much of their inspiration, was conceived because of the existence of printed maps, that did not live up to their objectives, and therefore clamoured for improvement.

Printing meant the democratization of access to maps. We are currently experiencing an information revolution which provides us all with computing power that we have never had at our disposal before. Our PCs and workstations now allow us to analyse large spatial data files at will. In the geographical information systems referred to here, these files can be combined and processed at enormous speed, and the results of these analyses can be portrayed on the monitor screens in map form in no time at all.

The danger here is that, as these GIS images might not be printed and multiplied, nor meant for a wider audience at all, one might fall back into the cartographic Middle Ages, i.e. to a period when, because of a lack of general access to manuscript maps, the discipline developed at a very slow pace, if it did so at all. This is because this restricted access meant restricted criticism and no urge for improvement. So the message is that even if the maps produced on GISs are ephemeral, the decisions made with these devices are so important that users should still take cognizance of the map design efficiency developed by cartographers and implement it in their work. This is a point to take note of, especially as the Internet, and in particular the World Wide Web, becomes a noticeboard and source for maps.

So this is what this book intends to provide: sufficient relevant knowledge of cartography for GIS users for the production and use of effective visualizations of spatial information.

It developed from a textbook (*Cartography: Design, Production and Use of Maps*) for cartography students in the Netherlands, published by Delft University Press in 1987. Its many illustrations were produced by practical cartographers from both Utrecht and Delft Universities. It was re-edited for its second edition in 1990 by describing and inserting the new automated processing of spatial data, with colour illustrations added, sponsored by the Netherlands Cartographic Society. This new edition published by Addison Wesley Longman has again been completely rewritten, now from the perspective of spatial information transfer, with visualization of spatial information as the central issue, and GIS users as the prime audience. In order to facilitate smoother production of the illustrations, these were all recreated on a computer or scanned in order to be available in digital form. This task fell to Axel Smits and Martin Jutte of Delft University. In rewriting it, we also changed the book's preoccupation with the situation in the Netherlands to a more international setting. The preference of one of the authors to spend his days of leisure in the Lake District made this all the more easy. Fate played into our hands by the fact that Maastricht municipality, when we had just developed a database for it as spin off for a consulting project, was selected as the site for signing treaties on the European Union. This new notoriety allowed us to use this example in our texts as well, and you will encounter it frequently.

Much to our regret we omitted the many references to non-English sources at the advice of our reviewer, as he/she made so much good sense in all the other points raised. In the Netherlands we are used to having access to cartographic literature from both France, Germany and Anglo-Saxon countries, and we thought that reflecting this would be to the benefit of our readers. So if you miss those references, we are not to be blamed. Lapses in the language would be the other item. Not being native speakers the decision to have Dutchmen write a textbook in English was probably soon regretted by the people at Addison Wesley Longman, but they never showed any sign of it.

This was the first book we wrote for which we had daily contact, if we wanted so, through e-mail with the publisher, and this speeded up the production enormously. Even so, though the last changes were made to the text in October 1995, some of the information is bound to be out-of-date by the time it will be available. But by commenting upon the newest developments we are confident that most of the information presented will outlast this century.

Menno-Jan Kraak
Ferjan Ormeling
October 1995

Acknowledgements

We are grateful to the following for permission to reproduce copyright material:

Fig. 1.10 sheet 612F (1954), 1.12 sheet 62A, (1954) and sheet 69B (1989), 2.5 TOP25 Raster, 5.21 sheet 69W (1991) and sheet 69B (1989), 5.35b sheet 37O (1993), 7.3 sheet 49G (1989) and colour Plate 13 TOP25 Raster 32C from © Topografische Dienst, Emmen; Fig. 2.4 published by permission of the Controller of HMSO and the UK Hydrographic Office © British Crown copyright 1995; Fig. 4.7 from Spot Management and decision making tool p34 & 35, © CNES (1986) - Spot Image distribution; Fig. 4.11 Maps 11 and 12 from *People in Britain,* Crown Copyright National Statistics; Fig. 5.27 Seite 15, Schweizer Weltatlas, p15, (1981) © Conference of the Cantonal Ministers of Education; Fig. 5.27b and colour Plate 12a (Grote Bosatlas 51st edition) and Plate 12b (Wolters Wereldatlas) from Wolters-Noordhoff Atlas Productions; Fig. 5.34a, 18-22 Tours 1:50.000 (1977) Institut Geographique National, The Netherlands; Fig. 5.34c (detailed Lanranger sheet 89 1:50.000) and 5.41 (detail leaflet Superplan) reproduced from Ordnance Survey with the permission of The Controller of Her Majesty's Stationery Office © Crown copyright; Fig. 5.34d sheet 2041 08/Ojakkala: 1:20.000 (1993), National Land Survey of Finland; Fig. 5.34e Charleston Quadrangle (S.C), 1:24.000 (1983), United States Geological Survey; Fig. 5.34f Karte 2515–1:25.000 (1988) reproduced by permission of the Swiss Federal Office of Topography, 18 Oct. 1995; Fig. 5.39 200 years Ordnance Survey Issue 17 September 1991 reproduced by permission of Royal Mail Stamps; Fig. 10.1b from Alexander Weltatlas, 1e Auflage Seite 71 Klett Perthes, Stuttgart (1989); Fig. 10.2a and b Atlas Van Nederland, 2nd edition p7, and 10.2c from ANWB wegenkaart (1993); Plate 14 Mindscape's Multimedia Atlas V.5, Mindscape (1994).

Cover image of Kilimanjaro adapted from *Diercke Atlas* p131, Westermann Schulbuchverlag GmbH; Kenya cover image from 'Kenya - a Geomedical Monograph', Heidelberger Akademie der Wissenschaften verlegten Buch (H.J. Diesfeld & H.K. Hecklau, 1977) in Serie Medizinische Landerkunde Band 5.

Whilst every effort has been made to trace the owners of copyright material, in a few cases this has proved impossible and we take this opportunity to offer our apologies to any copyright holders whose rights we may have unwittingly infringed.

CHAPTER 1

Geographic information systems and maps

1.1 The map as an interface to GIS

Maps have been used for centuries to visualize spatial data. They help their users to better understand spatial relationships. From maps, information on distances, directions, and area sizes can be retrieved, patterns revealed, and relations understood. During the last decade, developments around digital spatial data handling gained momentum. Consequently, the environment where maps are used has changed considerably for most users. With the computer came on-screen maps. Through these maps, the database from which they are generated can be queried, and some basic analytical functionality can now be accessed through menus or legends. In the 1980s these software packages that allowed for queries and analyses of spatial data became known as geographical information systems (GISs). As their functionality matured, their application spread to all disciplines working with spatial data. GIS introduced the integration of spatial data from different kinds of sources. Its functionality offers the ability to manipulate, analyse and visualize the combined data. Its users can link their application-based models to the data contained in the systems, and try to find answers to questions like, 'which is the most suitable location to start a new branch of a supermarket chain?', or 'what effect will this plan, or possibly its alternative, have on the surrounding area?'

Maps are no longer only the final products they used to be. The paper map functioned, and functions, as a medium for storage and presentation of spatial data. The introduction of on-screen maps and their corresponding databases resulted in a split between these functions (Figure 1.1). To cartographers it brought the availability of database technology and computer graphics techniques which resulted in new and alternative presentation options such as three-dimensional and animated maps. In a GIS environment, spatial analysis often begins with maps; maps support judging intermediate analysis results, as well as presenting final results. In other words, maps play a major role in the process of spatial analysis.

In a GIS environment visualization is applied in three different situations. First, visualization can be used to explore, for instance in order to play with unknown and often raw data. In several applications, such as those dealing with remote sensing data, there are abundant (temporal) data available. Questions such as 'What is the nature of the dataset?', or 'Which of those datasets reveals patterns related to the current problem studied?' have to be answered before the data can actually be used in a spatial analysis operation. Dataviewers (tools to graphically browse the database) play an important role here. Secondly, visualization is applied in analysis, for instance in order to manipulate known data. In a planning environment the nature of two separate data sets can be fully understood (e.g. the groundwater level and the possible location of a new road), but their relationship cannot. A spatial analysis operation, such as overlay, can combine both data sets to determine their possible spatial relationship. The result of the overlay operation could, when necessary, be used to adapt the plans. Thirdly, visualization is applied to present (e.g. to communicate knowledge of spatial information). The results of spatial analysis operations can be displayed in well-designed maps easily understood by a wide audience. The cartographic discipline offers design rules to do so.

Considering these three different fields of visualization in GIS (exploration analysis and presentation), it

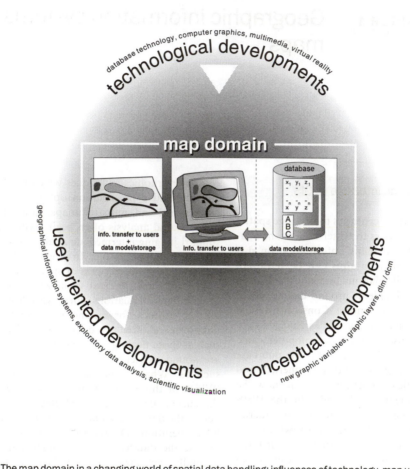

Figure 1.1 The map domain in a changing world of spatial data handling: influences of technology, map user environments and internal developments lead to a change in functions

can be noticed that the tools for presentation are the most highly developed (Robinson *et al.*, 1995). While producing maps to communicate spatial information, cartographic rules (together called 'cartographic gramar') are available to make the maps effective. However, as these rules are not part of the GIS software, GIS allows users to produce their own maps even when they are unaware of cartographic grammar. In other words, there is no guarantee that the maps will be effective. These cartographic rules could also be applied in the analysis phase, but the necessity to do so would be less strong here. When cartographers and analysts discuss this matter, the second group would always claim 'Who cares about your rules, as long as one understands one's own maps?' And because the analysts knew their own data they probably would understand their own maps, but when showing their maps to others trouble would start. In a data exploration environment it is likely that even the user does not know the exact nature of the data. It is not known how cartographic rules will function here. It is obvious that more appropriate visualization methods will have to be found for this situation.

At this moment the terms 'private visual thinking' and 'public visual communication' should be introduced (DiBiase, 1990). Private visual thinking refers to the situation where users work with their own data, and public visual communication refers to the situation of cartographers and their well-designed maps. The first describes the exploration circumstances and the second presentation circumstances. Analyses can be found somewhere in the middle along a line between the two. This becomes more evident when it is realized that private versus public map use (i.e. maps tailored to an individual versus those designed for a wide audience) is just one of the axes of the so-called map use cube, first introduced by MacEachren (1994). Along the two other axes the revelation of the unknown versus the presentation of the known, respectively high versus low interaction, are plotted. This is shown in Figure 1.2.

Most chapters in this book concentrate on maps that should communicate spatial information (the lower left front corner of the cube). However, recent developments in cartography and other disciplines handling spatial data not only require a new line of thought, they also create one. This can be illustrated by plotting the evolutionary stages of the development of electronic atlases in the cube along the diagonal from the corner 'wide audience, presenting knowns, and low interaction' towards the corner

'private use, presenting unknowns, and high interaction' (Figure 1.2b). Early electronic atlases were, in effect, sequential slide shows, but the more advanced electronic atlases have high interactive multimedia mapping capabilities, and allow users to combine their own data with atlas data. Each category of map use in the Figure 1.2 cube asks for its own visualization approach. New cartographic tools and rules have to be found for these approaches. They are probably not as restrictive as traditional cartographic rules, but on the other hand not as free as the technology allows either. Chapters 9 and 10 will concentrate on this new field.

The demand for sophisticated spatial data presentation is further stimulated by developments in scientific visualization, multimedia, virtual reality and exploratory data analysis. In each of these external developments influencing GIS and maps it would appear that from a technical point of view, there are almost no barriers left. The user is confronted with a screen with multiple windows displaying text, maps, and even video images supported by sound. Important questions remain. Can we manage all the information that reaches us? What will be the

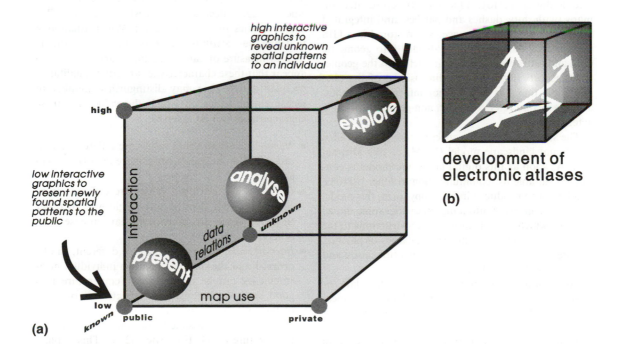

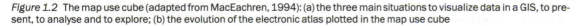

Figure 1.2 The map use cube (adapted from MacEachren, 1994): (a) the three main situations to visualize data in a GIS, to present, to analyse and to explore; (b) the evolution of the electronic atlas plotted in the map use cube

impact of these developments on the map in its function to explore, analyse and present spatial data? This book tries to provide an overview of the role maps will play both today and in the near future in the world of spatial data handling.

1.2 Spatial data

Geographical information systems are different from other information systems in that the data contained in them can, as a special characteristic, refer to objects or phenomena with a specific location in space and therefore have a spatial address. Because of this special characteristic the locations of the objects or phenomena can be visualized, and these visualizations – called maps – are the key to their further study. Figure 1.3 shows how objects from the real world that can be localized in space (like houses, roads, fields, or mountains) can be abstracted from the real world as a digital landscape model, according to some predetermined criteria, and stored in geographical information systems (as points, lines, areas or volumes) and later (after being converted into a digital cartographic model) represented on maps (with dots, dashes and patches) and integrated in people's ideas about space. When stored in a GIS these spatial data are usually divided into geometric data and attribute data. The first refers to the geometrical aspects (location and dimensions) of the phenomenon one has (geometric) information about, and the second refers to other, non-geometrical characteristics.

The stored data set of a specific study area is called the digital landscape model (DLM) of this abstraction. As soon as this digital landscape model is considered suitable for communication to other persons, and has to be produced in hardcopy form, this model has to be converted into a digital cartographic model (DCM), which consists of series of instructions to the plotter or printer, to produce dots, dashes or patches, in different sizes, colours, etc., for multiplication and distribution (Figure 1.3).

For data to qualify for the tag 'geometric data' or 'georeferenced data', information about their location would be required. This can be geographical or reference grid coordinates, code numbers that refer to statistical areas, topological terms (e.g. A is in between B and C), or nominal terms, as in street addresses. The spatial nature of the objects can be expressed in their shapes, i.e., the shape of the abstractions with which one represents objects from the real world. There is a basic subdivision into point-, line-, area- or volumetrically-shaped objects (see Figure 1.4), and this can be further subdivided into, for instance, elongated, triangular, irregular or convex shaped objects. In a sense, this is scale- or resolution-dependent, as a populated settlement will be rendered by a point in a national context, and as a built-up area in a municipal context.

Whether the objects or phenomena from the real world are abstracted as discrete or continuous is very important for subsequent storage and mapping procedures. Discrete objects can be bordered on all sides, and the coordinates of these borderlines can be made explicit. These can either be the locations of tactile objects (houses, streams, etc.) or of predetermined areas (states, enumeration areas or distribution areas). Continuous representations are abstractions of those phenomena that are considered to change non-incrementally in value. They can be tactile or measurable (like precipitation data, or gravity field data) or be based on models (like isochrones).

For later visualization procedures it is essential that the nature of the attribute information be established. These attributes can refer to visible characteristics (e.g. deciduous trees) and invisible characteristics (e.g. temperature). When attempting to define these attribute values of objects, one usually tries to measure or categorize them, and then it will appear that these characteristics are either qualitative or quantitative. One may distinguish a number of measurement scales on which the values for these characteristics can be assessed:

- *Nominal scale*: attribute values are different in nature, without one aspect being more important than another.
- *Ordinal scale*: attribute values are different from each other, but there is one single way to order them, as some are more important/intense than others.
- *Interval scale*: attribute values are different, can be ordered and the distance between individual measurements can be determined. Temperature is a good example: because the respective zero-points of their measurement scales have been selected at random, it is impossible to say that for instance a temperature of 64 °F is twice 32 °F. This is plain when the values are converted into Celsius and become 18 °C and 0 °C respectively.

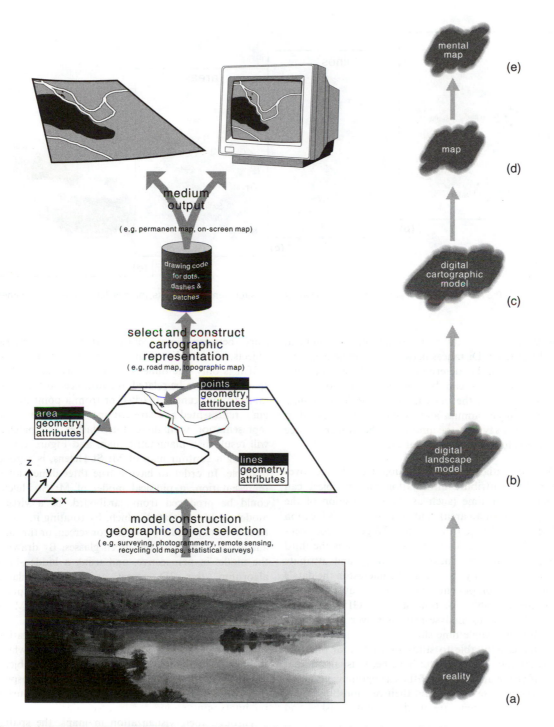

Figure 1.3 The nature of spatial data: from reality (a), via model construction and selection to a digital landscape model (b), followed by selection and construction to a cartographic representation towards a digital cartographic model (c), presented as a map (d), which results in the user's mental map (e)

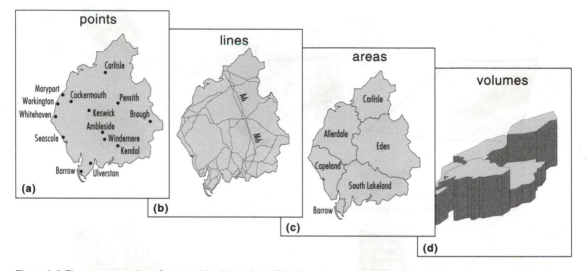

Figure 1.4 The representation of geographic objects in a (digital) environment as (a) points, (b) lines, (c) areas and (d) volumes

- *Ratio scale*: attribute values are different, and can be ordered. Distances between individual measurements can be determined, and these individual measurements can be related to each other. If, for instance, the per capita income in Sri Lanka is $300 per annum and in Bangladesh $150, then one can say that the amount in the former is twice the value of that in the latter.

All the spatial data will be subject to changes over time: the attribute information on an object can change over time (such as the composition of the population of an area), and even the object's location itself may change (for instance, Wegener's continental drift). The data's time stamp is seen as the third major component, next to geometry and attribute values. Especially these days the interest in the data's temporal component increases because of the expanded number of time series in GIS databases, and the wish to analyse processes over time instead of during a single time slice.

One is only able to study or analyse or interpret spatial data after the data have been visualized, that is, after the application of the cartographic grammar to render these objects and their relationships. Here, symbols and signs are used, i.e. dots, dashes and patches, and these can vary in size, shape, texture, colour, grey value and orientation (see Chapter 6). These signs are linked to the objects or relationships, and by doing so one is able to convey spatial relation-ships between point, object, area or volumetric objects, in a number of dimensions, to the map user.

If it is only one dimension that is available, then spatial data can be related, for instance, to their distance from a central market, or from a point of origin – represented as a straight line. Two-dimensional representation with these dots, dashes and patches will result in a planimetric map, and Figure 1.5(a), showing a contour map of Mt St Helens, is a good example. In order to have a true three-dimensional representation, a physical model of Mt St Helens could be produced from cardboard, or a virtual model could be created, which, by rotating it, could be seen from all sides on a monitor screen, or through anaglyphs, using red and green glasses. By drawing the model in perspective, and using a hidden-lines algorithm, this 3D aspect could be simulated (represented in Figure 1.5b). The current description of this type of rendering is '$2\frac{1}{2}$ dimensional'. Maps with hill-shading are another example of this.

If one adds the time dimension, the representation will become four-dimensional (Figure 1.5c), when, through juxtaposition of two states of this object, for instance on 17 May and 19 May 1980, respectively, the change in its geometry or attributes during the intervening period can be ascertained.

Through their visualization in maps, the spatial relationships of objects can be made visible. These spatial relationships will usually refer to relations to some specific location on the earth's surface, and

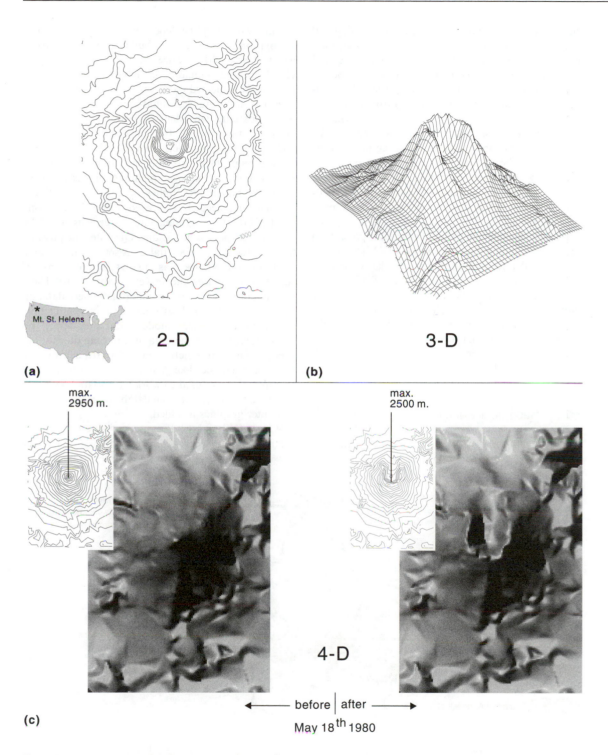

2-D

(a)

3-D

(b)

max.
2950 m.

max.
2500 m.

4-D

(c)

before | after
May 18 th 1980

Figure 1.5 The dimensionality of geographic objects: (a) 1D (inset map) and 2D; (b) 3D; (c) 4D/time

these relations can have many forms. Primary spatial relationships are those between objects and their location on earth, or between these objects and their attributes, such as the type of vegetation occurring at that location, or the type of road classification. By visualizing object categories from a data file (e.g. car factory locations or stream networks, or fields used for horticulture), relations between the elements for an object category will be made clear, and one will be able to perceive patterns or spatial trends. By combining geometric and attribute data, one would be able to perceive how the locations of elements from different object categories might influence each other.

Next to these primary relationships it is possible to perceive secondary types: relations of objects to linear or areal reference units, such as that of inhabitants to surface area, or the number of cars to the length of the highway network, or the relative amount of horticulture in all agricultural areas. One could go further and introduce other dimensions (like height or time), so that tertiary or even higher order relationships would emerge.

1.3 Geographical information systems

For most disciplines working with spatial data, one of the first uses of the computer was to create an inventory of discipline-dependent data. In this period cartographers worked to build a database from which they could produce the maps that were previously created manually. In a next phase spatial analysis of the collected data was emphasized. Forestry scientists, for instance, would apply statistical methods to spatially dependent attribute data. For cartographers this meant the possibility of creating different derived products from the existing database. Nowadays problems are approached in an interdisciplinary way. In physical planning processes or in environmental impact studies, data from many different fields are needed. This need led to the development of GIS. Cartographic knowledge is used in GIS to create proper visualizations. GIS offer the possibility of integration of spatial data sets from different kinds of sources, such as surveys, remote sensing, statistical databases and recycled paper maps. Their functionality allows one to manipulate these data, or to set up spatial analysis operations in conjunction with application-based models, and they allow for the visualization of the data at any time during this process. The core functionality of a GIS is provided by those disciplines like geography, geodesy and cartography that are used to work with spatial data. To this core, functionality from database technology and computer graphics is added.

Why is GIS unique? Because it is able to combine spatial and non-spatial data from different data sets in a spatial analysis operation in order to answer all

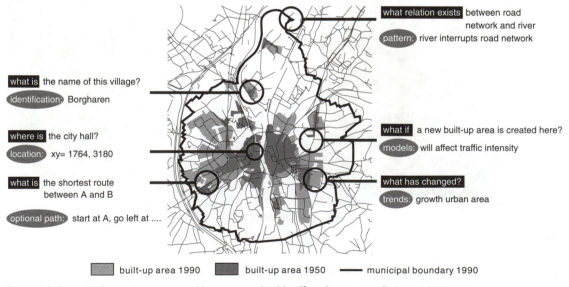

what is the name of this village?
identification: Borgharen

where is the city hall?
location: xy= 1764, 3180

what is the shortest route between A and B
optional path: start at A, go left at ....

what relation exists between road network and river
pattern: river interrupts road network

what if a new built-up area is created here?
models: will affect traffic intensity

what has changed?
trends: growth urban area

built-up area 1990   built-up area 1950   — municipal boundary 1990

Figure 1.6 Typical GIS questions answered by maps used to identify, to locate, or to find spatial patterns

kind of questions. Traditional statistical packages or basic CAD packages could only deal with one of those components or with a single data set. What type of questions can be answered by a GIS (Figure 1.6)?

- What is there . . . ? *Identification*: by pointing at a location on a map, a name, or any other information stored on the object, is returned. This could also be done without maps, by providing the coordinates, but this would be far less effective and efficient.
- Where is . . . ? *Location*: this question results in one or more locations that adhere to the criteria of the question's conditions. This could be a set of coordinates or a map that shows the location of a specific object, or all buildings in use by a certain company.
- What has changed since . . . ? *Trends*: this question includes spatial data's temporal component. A question related to urban growth could result in a map showing those neighbourhoods built between 1950 and 1990.
- What is the best route between . . . ? *Optimal path*: based on a network of paths (e.g. roads or a sewage system), answers to such queries for the shortest or cheapest route are provided.
- What relation exists between . . . ? *Patterns*: questions like this are more complex and often involve several data sets. Answers could, for instance, reveal the relationship between the local microclimate and location of factories and the social structure of surrounding neighbourhoods.
- What if . . . ? *Models*: these questions are related to planning and forecasting activities. An example is: what will be the need to adapt local public transport network and its capacity when a new neighbourhood is built north of the town?

Of course, one does not only query a GIS; one also uses it interactively, for instance in physical planning procedure, through manipulation of designs, etc.

GIS development was stimulated by individual fields such as forestry, defense, cadastre, utilities and regional planning (see Chapter 4). Since they all have different backgrounds and different needs, the functionality of the software GIS initially used was different as well. It ranged from statistical analysis packages to computer-aided design packages. Functionality was added and each of these groups started to call their software a GIS package. This resulted in different meanings for the same term.

Next to GIS, literature offers wordings such as land information systems, geo-base information systems, natural resources information systems and geo-data systems (Maguire *et al*. 1991). Figure 1.7 shows the relation between the most common terms. Irrespective of the terminology, it is obvious (Maguire *et al*. 1991) that three views on GIS exist: from a spatial analysis perspective, from a database management perspective, and from a map production perspective. The first attracts most supporters. The last view is preferred by those organizations involved in the 'production' of topographic data.

The multidisciplinary background of GIS led to a multitude of definitions. In general they can be split in two groups: those with a technological perspective and those with an institutional/organizational perspective. An example of the first is the definition by Burrough (1986): 'a powerful set of tools for collecting, storing, retrieving at will, transforming and displaying spatial data from the real world'. An example of the second that of Cowen (1988): 'a decision support system involving the interaction of spatially referenced data in a problem-solving environment'. So it is the potential combination of different data sets that is paramount. A working definition for this book is derived from a combination of the two above: a computer-assisted information system to collect, store, manipulate and display spatial data within the context of an organization, with the purpose of functioning as a decision support system.

In order to manipulate spatial data, to procure added value, a GIS consists of software, hardware, spatial data and people (the organization). These components communicate via a set of procedures. In Figure 1.8, which summarizes the view of GIS adapted in this book, the scheme in the centre presents the structure of GIS. It is made up from two nested boxes. The centre shows the components of a traditional information system (input, manipulation and storage), all oriented on spatial data, the unique and basic ingredients of the system. The outer box represents the organization in which GIS functions. The configuration of the scheme stresses the need for a proper user interface and management of the system.

Each organization will require a GIS with emphasis on a specific set of functions. In general, functions are needed for data input and encoding (e.g. digitizing, data validation and structuring options), data manipulation (e.g. data structure and geometric conversions, generalization and classification options), data retrieval (e.g. selection, spatial and statistical

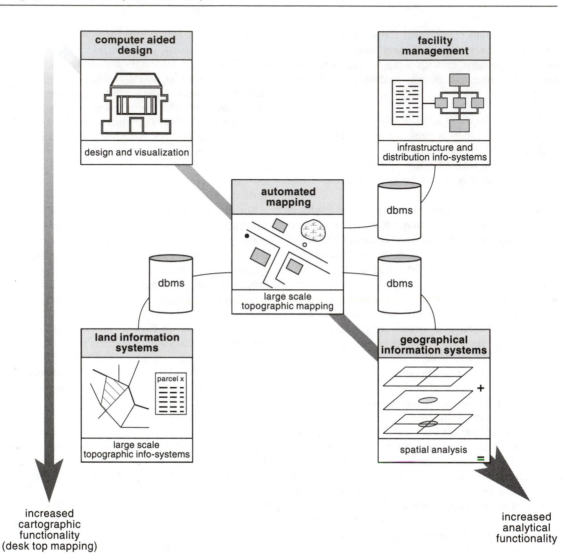

Figure 1.7 Relations between several types of spatial information system. Down the scheme the cartographic functionality increases, while from left to right the analytical functionality improves, which is closely related to the use of database management systems (dbms)

analysis options), data presentation (mainly graphical display options) and integrated data management.

1.4   GIS issues

Spatial analysis operations are the new items GIS has to offer to the spatial data handling community. However, its principles have been known since the 1950s from fields like quantitative and statistical geography. This section will explain the principles of these operations, illustrated with a simple example that demonstrates the strength of GIS and the role of maps (Figure 1.9). The example deals with an issue in the Netherlands municipality of Maastricht.

The first step in a spatial analysis operation should be the definition of its objective and the conditions to adhere to. These conditions can include specific restrictions and constraints. In the Maastricht case

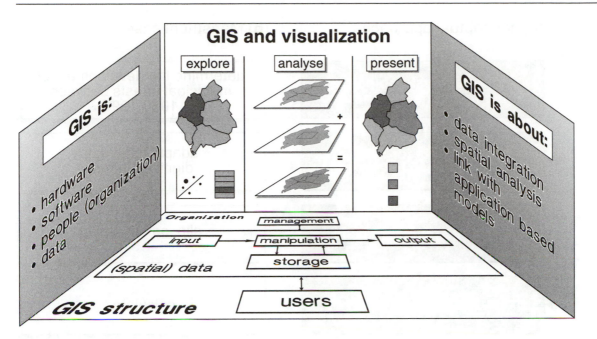

Figure 1.8 View on GIS: its structure and relation to visualization

the municipal authorities wanted to know how the municipal forests had developed between 1950 and 1990. They wanted to have a map indicating obliterated, unchanged and new forests, as well as a table with the size of the forest parcels. One of the reasons for this analysis was to check whether a large private company in the municipality had adhered to the conditions agreed upon. The company had a concession to excavate marl in the south of the municipality, but it had to plant new trees in those areas that had been affected by its operations.

The second step in an analysis is to prepare the spatial data. Usually not all data are available in a format that fits the requirements of the spatial analysis. Some data have still to be collected in the field, other data must be bought from external sources or other municipal departments, while there is a fair chance some of the data are available on paper maps or in tables only. When available digitally, they could still be in a different coordinate system. Other problems that are likely to occur are the data being available on a different aggregation level or the density of coordinates in one set being too low to be compared with another data set.

An example of the first of the problems are statistics being available at a neighbourhood level instead of at a street block level. The main GIS task at this stage is to make sure all data can be integrated and formatted so that they will be fit for use.

To answer their query, Maastricht municipality needed data on land use from both 1950 and 1990 to extract information on the forest parcels. The 1990 data were obtained from a database that contained data from the topographic map 1:25 000. The 1990 municipal boundary was taken from the large-scale map database available. However, it was too detailed to be used in conjunction with the 1990 topographic map data. It had to be generalized first. The 1950 data were not available in a digital form. The original topographic map had to be ordered from the municipal archives, and had to be converted into digital form. An interesting problem that occurred during the conversion was that part of the area for which the data were needed, within the 1990 municipal boundary, was not on the 1950 map, because the area of Maastricht municipality had been enlarged considerably since, as can be seen in Figure 1.10(a). The generalized 1990 boundary was used as a digitizing mask to extract the correct information from the 1950 map (Figure 1.10b). The same 1990 boundary was used to select the land use data from the 1990 topographic map database. This database contained, in several layers, all data of a topographic map sheet that covers a large area in the southern part of the

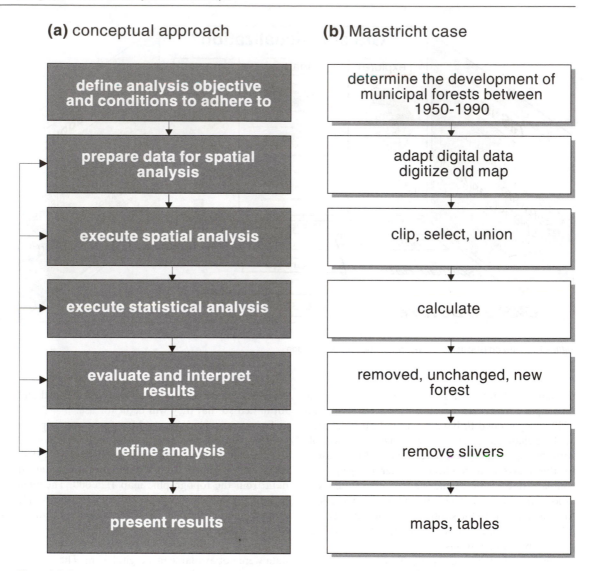

(a) conceptual approach

(b) Maastricht case

define analysis objective and conditions to adhere to	determine the development of municipal forests between 1950-1990
prepare data for spatial analysis	adapt digital data digitize old map
execute spatial analysis	clip, select, union
execute statistical analysis	calculate
evaluate and interpret results	removed, unchanged, new forest
refine analysis	remove slivers
present results	maps, tables

Figure 1.9 Spatial analysis: (a) conceptual approach; (b) Maastricht case

Netherlands. From it, only data related to Maastricht were needed. Working with the whole database would slow down the execution of the operation because of the large amount of data.

The third step in the operation was the actual execution of the spatial analysis. Most packages have all kinds of operations available, which somehow operate on the spatial and the non-spatial component of the data. In general, one distinguishes three major types of spatial analysis operations: overlay and buffer operations, network operations, and surface operations. The first category often combines

several data sets based on certain criteria, the second category uses an infrastructure network to find optimal paths, while the third category determines all kind of (terrain) surface characteristics. Most packages allow the combination of these operations. The Maastricht case was limited to a simple overlay operation. The data sets from 1950 and 1990 were combined, which revealed those forest parcels unchanged, those removed and the new ones. Figure 1.11 demonstrates this process. From the 1950 and 1990 land use data sets the forests were selected to create two new data sets, forests 1950

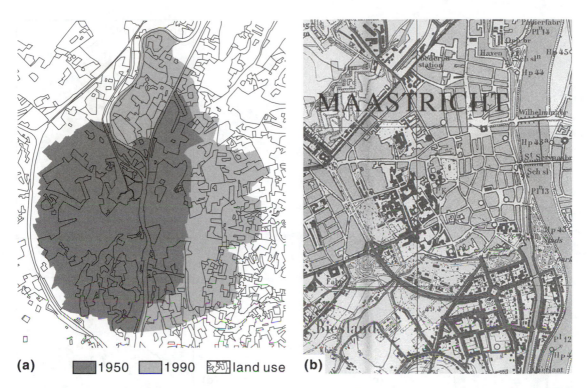

(a) ▮▮1950 ▭▭1990 🔀land use   **(b)**

Figure 1.10 Some of the available data in the Maastricht case: (a) land use derived from a topographic database – the 1950 and 1990 municipal boundaries are given for reference; (b) a detail of the 1950 topographic map to be digitized (sheet 61F; Courtesy Topografische Dienst Nederland)

and 1990. It was those two sets that were combined in the overlay operation. This calculated all possible intersections between the 1950 and 1990 forest parcels. In this process all attributes were inherited and saved in the final data set. Based on these attributes, the map in Figure 1.11(e) could be drawn. The command 'draw all forest parcels with an attribute year 1950 and not year 1990' would result in all the parcels obliterated since 1950 being drawn.

The statistical analysis of the results is the step next to the spatial analysis. It is executed to fulfill the conditions set when the objective was determined. When, in the Maastricht case, one criterion would have been that the municipality is only interested in those parcels over a certain area or perimeter or when, in addition, it would like to know the average size of the new parcels, some basic statistics could be applied here. In complex spatial analysis operations it would be likely that more sophisticated statistical methods are needed. The next step in the analysis was the evaluation and interpretation of the results. In general, when executing a spatial analysis one has

certain expectations when there is some familiarity with the data. If the map revealed large new forest parcels in the city centre it would be safe to conclude that something went wrong. If this was the case one would have to correct certain steps in the analysis or refine analysis conditions.

Looking at the result in the map in Figure 1.11(e), everything seems to be right. However, if one takes a closer look at the result it can be seen that the result is not quite perfect. Figure 1.12(a) shows why. It reveals lots of small polygons indicating change. However, if the enlargement in the same figure, together with the two map details, are analysed (Figures 1.12 a–c) it is obvious that in reality nothing has changed at all, although the GIS operation created 11 new polygons. These polygons are called sliver polygons. A comparison of the basic statistics of the resultant data set with the original data set would have caused suspicion as well. The original data sets have 32 and 36 polygons respectively, while the new data set has 100 mainly small polygons. The main reason for their occurrence is that the same feature

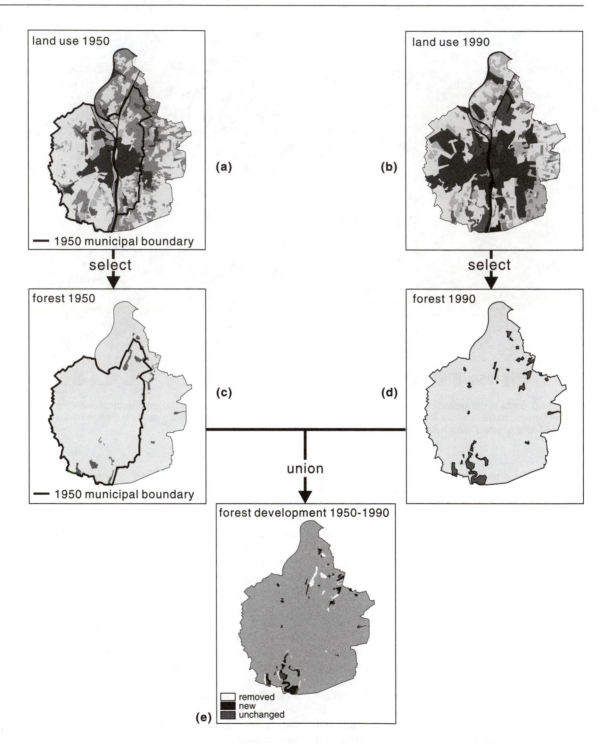

Figure 1.11 An overlay operation: (a) and (b) show the basic land use data from 1950 and 1990; (c) and (d) show the selected forest parcels from 1950 and 1990; (e) the overlay result: removed, unchanged and new forest parcels

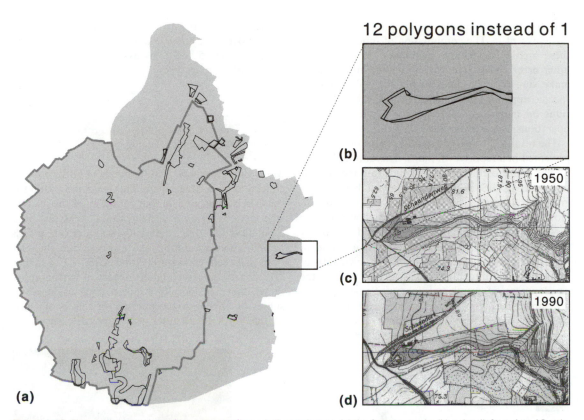

12 polygons instead of 1

Figure 1.12 Figure 11.1(e) enlarged: overlay results and sliver polygons: (a) the forest parcels; (b) a detail: from 1 to 12 polygons; (c) and (d) the original 1950 and the original 1990 parcels (sheet 62A and 69B respectively; Courtesy Topografische Dienst Nederland)

in both the 1950 and 1990 data sets has a different geometry. The digitizing of the 1950 and the 1990 maps was not performed by the same operator. Even if it would have been the same operator, it would have been unlikely that the same points would have been selected during both digitizing sessions. For a problem like this most GIS software offers simple solutions. One can delete all polygons with a size smaller than a certain threshold or calculate an average polygon boundary. Both approaches have disadvantages. But if the sliver polygons are not removed and the results are used in future spatial analysis, the errors will propagate into the future results.

During the collection of spatial information many types of error can be made: errors in measuring, classifying or categorizing data, localization errors, mistakes in data entry, etc. When these data are not directly incorporated into a GIS during the collection process, but are, for instance, mapped first because the new technology had not been applied yet, then other kinds of errors will emerge. Among these are generalization errors or misrepresentations due to data amalgamation, reproduction errors and errors due to deformation of the printing paper. When these map data are subsequently digitized or scanned for input into the database, these errors are at least duplicated in it, but more probably the digitizing process itself will be another source of error as the Maastricht case demonstrates. At this moment, it is not quite clear, however, how these errors (for error propagation, see Goodchild and Gopal, 1992) may affect the spatial analysis results, i.e. whether they would lead to uncertainties in the results of analysis operations that would exceed some critical level. There are not only errors in the input values; errors can also be caused by analysis operations themselves and by the application-based models used. Examples are spatial computational modelling techniques that forecast groundwater flow or polluted air diffusion and try to approximate reality but might in fact misrepresent it. The combination of input error and these spatial modelling techniques might lead to other error types. It is therefore very important to make sure that the data quality (i.e. suitability for specific applications) is sufficient before basing decisions on maps that represent the results of spatial operations executed on these data.

The results of a spatial analysis operation are often presented in a report with maps, diagrams and tables to emphasize certain points or illustrate the conclusions. Most GIS packages do have a basic cartographic functionality to create the graphics. However, dedicated desktop packages have a more extended cartographic functionality and are better suited to produce the final maps. An example of these are the maps created by the municipality of Maastricht for illustrating their final report on the development of municipal forests. Some of these are shown in Figure 1.13. Included is a qualitative map that shows forest developments as well as a

shaded relief map with the forest development map draped over it to show relations with the terrain surface. Chapters 5 and 7 discuss the characteristics of most existing map types.

1.5   The relation between GIS and cartography

Many of the concepts and functions of GIS were first conceived by cartographers. This is not only valid for the GIS output module, but for many of the processing actions (e.g. transformations, analyses) and input functions (e.g. digitizing, scanning) of a GIS as well. There are conflicting views regarding the relations between cartography and GIS, viz. whether GIS is a technical-analytical subset of cartography, or whether cartography is just a data visualization subset of GIS. For the purpose of this book, also written for GIS analysts who have to learn to use

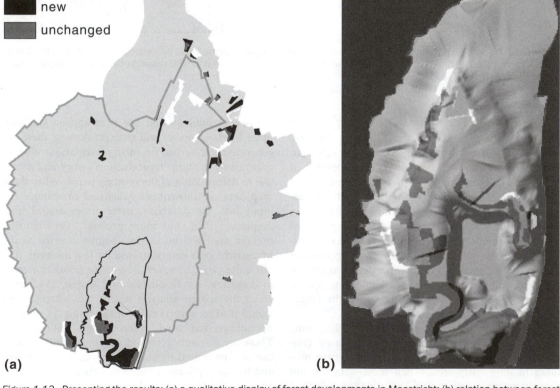

Figure 1.13 Presenting the results: (a) a qualitative display of forest developments in Maastricht; (b) relation between forest developments and the terrain

the cartographic method, cartography will be regarded as an essential support for nearly all aspects of handling geographical information for the following reasons.

- maps are a direct and interactive interface to GISs, a sort of graphical user interface with a spatial dimension;
- maps can be used as visual indexes to phenomena or objects that are contained in the information systems;
- maps, as forms of visualization, can both help in the visual exploration of data sets (also the discovery of patterns and correlations), and in the visual communication of the results of the data set exploration in GISs;
- in the output phase, the interactive design software of desktop cartography is superior to the output functions of current GISs.

These should be enough reasons for cartography to have an important place in GIS, but there are more reasons, if one looks at the context in which GISs are being used: they are aimed at decision support, and as this regards decisions about geographical objects, it should be visual decision support, in order to take into account the spatial dimension as well. In order to correctly use these visual decision support aids (the maps visualized on the computer screen or the hard copy output of these systems), the users should adhere to proper map use strategies (see Chapter 10). This ability to work with maps and to correctly analyse and interpret them is one very important aspect of GIS use. Strangely enough, not one GIS manual gives any clarification in this field, assuming that all GIS users are aware of the ins and outs of map use.

But there is another important decision support aspect to the information that is processed in and presented by a GIS: data quality. GISs are very good in combining data sets; notwithstanding the fact that these data sets might refer to different survey dates, different degrees of spatial resolution, or might even be conceptually unfit for combination, the software combine them and present the results. Cartographers, in compiling maps, have worked with different data sets for centuries, and have some experience in the transformations that are necessary in order to combine data sets with different resolution, projection, reference system, geoids, and dates of survey. They have developed transformations and modelling procedures (such as generalization) that

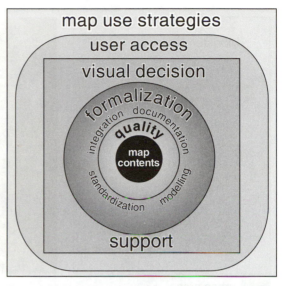

Figure 1.14 Visual decision support for spatio-temporal data handling. Key words in the GIS-cartography approach are map use strategies (how people make decisions based on maps), public access (how people work with the information), visual decision support (what the quality of the information is like), formalization (building expert systems). Based on Kraak *et al.* (1995)

take account of these differences and will allow for real data integration. They have developed documentation techniques that will describe all relevant data characteristics (meta-information) necessary for proper integration to take place. They have also lobbied for decades to standardize these documentation techniques so that the data sets can be easily exchanged (Figure 1.14). So much of the methodology for the determination of data quality is potentially available for GIS users from cartography.

As to the assessment of data quality, this can be defined, as was shown in the preceding section, as a measure for the suitability of data for specific applications. So one can, for instance, determine the precision x to which objects (such as parcels) in a data set have been localized, as well as the probability p with which these objects have been correctly classified or categorized (e.g. regarding their land cover). Now the GIS will allow the combination of this land cover information (e.g. surveyed in 1990) with precipitation information (e.g. surveyed for the period 1930–60, for five points in the area, and interpolated for all other locations), with a planimetric accuracy y of the rain gauge locations and a representativity factor z_{1-5} of these measurements for their surrounding areas. Well, what will be the value of correlations

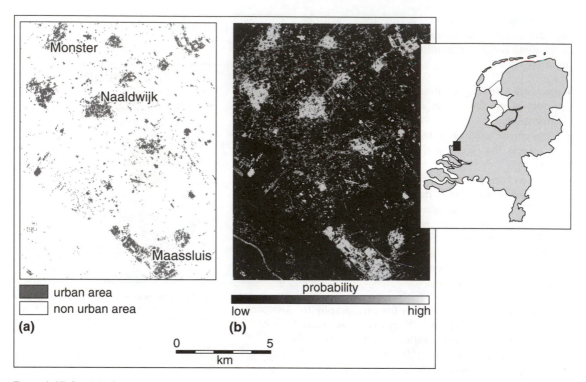

Figure 1.15 Spatial information and meta-information: (a) distribution of urban areas over a part of the western Netherlands; (b) probability map for the classification 'urban' (from Van der Wel and Hootsmans, 1993)

between precipitation data and land cover data, taking into account these accuracy, probability and representativity values? Until recently these have been disregarded in GISs, but in a mature GIS that really functions as a decision support system, these values should be indicated, to properly inform the decision-maker.

Figure 1.15 shows the classification of remote sensing imagery of urban areas in a part of the western Netherlands (near The Hague); the image on the left is a classification based on spectral qualities, the image on the right is a probability map for the classification 'urban'. It takes into account the potential confusion with related spectral signatures. The probability of a pixel being correctly classified as 'urban', and not as 'hothouse' or industrial complex or beach or bare soil, can be computed, and visualized probability values for correct classification like these should form an essential element of the decision-making process which takes place everywhere where spatial information is involved.

So GIS users can be provided with essential tools in all phases of collecting, processing and analysing spatial data, and communicating it to decision-makers. Those GIS users able to use maps are provided with the conceptual infrastructure for a correct decision-making procedure, and with the necessary information (meta-information) on the quality of the data contained in those maps.

1.6 Examples of the application of the cartographic method

It is possible to do without maps in a GIS: one could ask for the coordinates or addresses of sites that answer a number of requirements without ever visualizing them. But by doing so, one would deny oneself the opportunity of obtaining additional information from the process, as any spatial trends or patterns in the answer might never come to light. The information transfer without maps (e.g. through tables) would be more cumbersome as well. It would be wise, therefore, to apply the 'cartographic method', that is to visualize the spatial relationships between the objects using abstraction techniques and the transformation based on the graphical grammar explained in Chapter 6, in other words, to map it. Observing the spatial connections, relationships and patterns is only possible through the abstracting capacity of maps. As an example, two practical case studies will follow, showing the role that maps can play.

1.6.1   The location of the TGV or high-speed train in the Netherlands 'Randstad' area

It was decided in 1994 to extend the Paris–Brussels TGV link to Amsterdam. Consequently, as the existing rail links are already overburdened with traffic, and as extra foundations must be constructed because of its high speed, a new route for the rail link has to be selected, though the 'Green Heart' of the Randstad area, i.e. the non-urbanized centre of the urban agglomeration in the western Netherlands. The proposed route should spare the environment as much as possible (nesting birds should be disturbed as little as possible, and there should be no polluting influence on groundwater or on vegetation). Moreover, no valuable geo-scientific monuments should be affected.

An environmental information system (EIS) built for the Netherlands was therefore consulted. This EIS contains data on soils, groundwater, vegetation and fauna and even on rare geological outcrops (geoscientific monuments). These data have been collected for the EIS on a grid-cell basis. For each grid cell (1 by 1 km), dominant soil types (by putting the grid over a soil map), dominant vegetation types, the number of different vegetation types found per grid cell and the total number of vegetation type units, the types of wild animals that occurred, etc., were ascertained. Because of the fact that this information was stored in the EIS, the effect of the proposed routes could be easily estimated. In order to select the best route from a number of alternatives (Figure 1.16a), the susceptibility of the soils to water-table lowering (Figure 1.16b), the susceptibility of mammals to fragmentation of their habitat (Figure 1.16c) and the effects of disturbances and pollution on bird life (Figure 1.16d) for all the affected grid cells was determined. Subsequently the computer was used to calculate how many of these grid cells would be affected, and to what degree, for every proposed route. In other words, one could use the computer to define

the sum of the environmental values which, because of the construction of the TGV along the various routes, would be affected or nullified. This created the opportunity to select the route that would create the least damage.

1.6.2   Drawing up a weather forecast

Simultaneously, at fixed times, meteorological observations are made all over the world. These observations, expressed in numbers and presented in a predetermined sequence, are exchanged internationally. Thus each national meteorological institute has recent data available for an extensive area around it. The data (atmospheric pressure, temperature, wind velocity and direction, degree of cloudiness, precipitation, etc.) are subsequently plotted on a map (or as a weather chart, as seen for instance in Figure 1.17c). Since the data are collected and broadcasted every 3–6 hours, series of charts are created, which show the successive statements about the situation of the weather at the moment the recording was made. Now it can be seen on the chart in Figure 1.17(c) that it is raining, and that there is a wind blowing in an easterly direction. So it is to be expected that in the following hours the precipitation area will move in an easterly direction. From this one chart (Figure 1.17c) it may also be possible to make a forecast about the speed at which the precipitation area will move (since wind velocity data are also inscribed in the chart), but it cannot be guaranteed that the direction of the wind will not change. This can only be deduced from a series of weather charts, in which the one just mentioned is the most recent one. Only by comparing a number of successive weather images (e.g. Figure 1.17a–c), are meteorologists able to forecast with some degree of reliability the expected change in the weather and the speed at which the change will occur. On the basis of the trends and patterns that the meteorologists perceive from the succession of charts, a chart with a weather forecast for a moment x hours after the last observation can be produced by them (Figure 1.17d).

As well as the environmental values for each grid cell in the TGV example, all the weather data for the locations taken into account are contained in information systems. These meteorological information systems will be able to compare the actual weather situation to similar situations in the past, or to meteorological models of circulation types, and indicate what will usually happen in such a situation. So

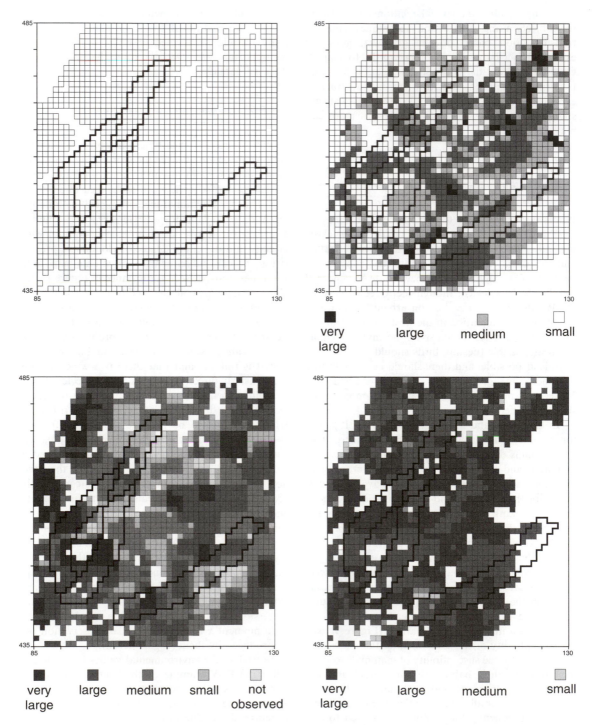

Figure 1.16 The location of the TGV or high speed train in the western part of the Netherlands: (a) alternative routes; (b) susceptibility to water-table lowering; (c) susceptibility of mammals to habitat fragmentation; (d) susceptibility of birds to traffic intensification

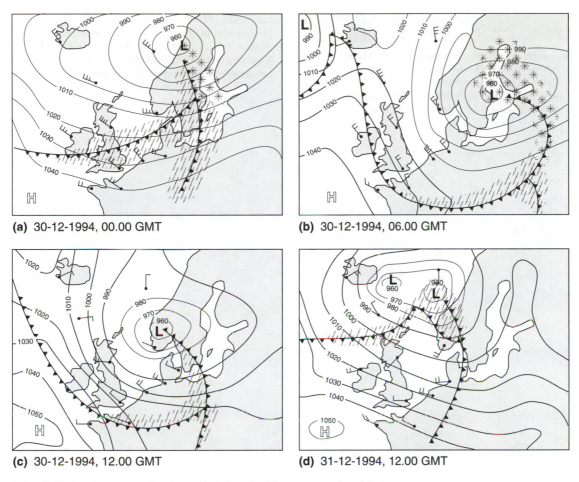

(a) 30-12-1994, 00.00 GMT

(b) 30-12-1994, 06.00 GMT

(c) 30-12-1994, 12.00 GMT

(d) 31-12-1994, 12.00 GMT

Figure 1.17 Drawing up a weather forecast (d), based on the trend seen from (a) – (c)

these GISs combine the memory and the computing power with the potential for analysis, and the system comes up with a number of potential developments, and even computes the probability of any of these developments being realized. Just as for the TGV route, it would be up to the expert working with the system to decide on the basis of these visualized spatial data, combining these spatial patterns with his or her knowledge of the area and the phenomenon.

But the cartographic method has applications outside of geographical information systems as well. As will be discussed in Chapter 9, digital maps are not only used in geographical information systems, but have a life of their own as well. Electronic atlases are one example. Also called electronic atlas information systems, their function is less one of information pro-

cessing than of answering specific questions, providing the support to integrate the answers in the mental map of the atlas user. This requires specific scenarios for a gradual immersion of the user into the new information environment. These atlas information systems can be extended to contain drawings, photographs, text and sound, and so become multimedia systems.

GISs are not yet well equipped enough to handle multi-temporal information, and it is here that animated cartography comes in. Animation techniques are being developed that show the spatial effects of developments at every stage. This presents extra potential for analysis, and is one of the avenues for advanced data exploration that will, in the future, also be available in a GIS.

Further reading

Burrough, P.A. (1986) *Principles of Geographical Information Systems for land resources assessment.* Oxford: Oxford University Press.

Goodchild, M. and S. Gopal (eds) (1989) *Accuracy of spatial databases.* London: Taylor & Francis.

Maguire, D.J., M.F. Goodchild and D. Rhind (1991) *Geographical Information Systems.* London: Longman.

Robinson, A.H., J.L. Morrison, P.C. Muehrcke, A.J. Kimerling and S.C. Guptill (1995) *Elements of cartography* (6th edn). New York: Wiley Inc.

CHAPTER 2

Data acquisition

2.1 The need to know acquisition methods

In geographical information systems it is usual for many files to have been combined, in order to boost the potential for analysis of the spatial data. In an ideal situation, all the data combined will have been collected, identified and measured on the same date, with the same spatial resolution, according to identical procedures, and consecutively entered into the GIS using the same method. It is only then that users can be sure of an adequate quality of the results of the analytical operations for which the files are being combined.

In practice, however, data acquisition is far from ideal: data are collected at different moments, are valid for different spans of time, have a different spatial resolution; and some might be collected in the field while others were taken from existing maps that were generalized to an unknown degree. Some files might have been entered after they have been made compatible using some rubber-sheeting technique, others may have been subjected to numerical transformation from other projections. Some might be based on random samples, others on complete surveys. In the case of numerical data collected at regular sample points, some might have been interpolated on the basis of linear, others on the basis of geometric types of progression.

So at least the potential situation exists that the data, when compatible at face view, might not warrant the conclusions drawn from their analysis. It could be the case that the analyst or the GIS user in general should be warned about the results, i.e. that the results should be interpreted with care. Traditional topographic maps used additional methods to enable the reader to assess data quality aspects, such as reliability diagrams to show that navigation in certain areas on the basis of the map would be hazardous, as the producers could neither guarantee the accuracy nor the completeness of the data. In the more complex world of GIS one needs numerical aids to indicate the quality of the data files, in order to be able to decide on the validity of analysis results.

It should be kept in mind that in all these cases one is collecting coordinates with which to describe locations of objects, with attributes the nature of which is determined either at the same time (e.g. during terrestrial topographic surveys) or later, in a lab (e.g. for soil surveys), or through field checking (e.g. for remotely sensed data).

The various spatial data acquisition methods for GIS can be divided into the following types (Figure 2.1).

- *Terrestrial surveys* Large-scale topographic data can be acquired through terrestrial surveys. Increasingly, such surveys lead to digital files that can be immediately imported into a GIS. When surveying new extensions for telephone companies or cable companies that have digital files of their networks at their disposal, surveyors would use the new 'total' stations with which the survey data are immediately edited and transformed into files that are extensions of existing files, so that the new data can be added to the existing data.

- *Photogrammetrical surveys* From aerial photographs object coordinates can be determined in the present analogue or (increasingly) digital stereoplotters, and imported directly into information systems. The attribute information required can be determined either through interpretations, or through field-checking.

Figure 2.1 Various types of data acquisition method: (a) surveying; (b) enquiries and statistics; (c) photogrammetry; (d) remote sensing; (e) digitizing maps; (f) census data

- *Satellite data*   Satellites have been built that contain scanners with sensors susceptible to the radiation emitted or reflected by the earth's surface. These scanners operate in such a way that the sensors measure radiation sequentially from patches or grid cells along paths perpendicular to the line of flight. These radiation data are later put in their proper spatial relationship and by doing so, a map is simulated. Here, data accuracy also depends on a number of correction techniques, for both the radiation values and the geometric accuracy. After these corrections the data frequently have to be resampled again in order to fit specific grids, or to be comparable to other data sets. By being collected for grid cells (with, for instance, 20 × 20 or 30 × 30 m resolution) the data are generalized from the start.
- *GPS data*   A special set of satellite data is provided by the Global Positioning System which, on the basis of 24 satellites, is able to pinpoint one's position three-dimensionally with an accuracy of a few centimetres. These GPS recordings are used to increase the accuracy of existing geo-referencing methods, or can be used directly in data surveys; they can be used for both point and linear surveying. The data are recorded on the basis of a global reference system, which can be transformed to local reference systems.
- *Digitizing or scanning analogue maps*   Manual digitizing refers to the registration with a cursor of sequences of characteristic points belonging to lines on a map, through which action the coordinates of the positions touched are recorded digitally. Scanning works as in a fax machine: optical records of the existence of specific colours at specific positions are transformed into files with information on positions with attributes (hue or colour value). If digitizing or scanning could be effectuated with a 100% correctness, the results would still depend on the accuracy of the original maps.
- *Using existing boundary files*   Commercially available boundary files (digital geometric descriptions of administrative units) can be acquired for applications that occur frequently or for a large number of uses: topographical files, files for car navigation, or boundary files to be used in conjunction with statistical data (such as for marketing applications). For these existing files it is essential that they are compatible with one's software, and therefore are based on the same standards as used by the buyer.

- *Socio-economic statistical files*   National statistical services are increasingly publishing their data on CD-ROM, with the appropriate software to query the files, or even to visualize them in map form (e.g. *Atlas de France*). These files are usually presented in standard formats that make them compatible with most current mapping packages. One of the items contained in these socio-economic data files is the standard area codes which relate the data to the areas for which they have been collected. This link between data and area should be preserved when entering the data into other packages. It is through these code links that these packages will know what data to link to specific areas.
- *(Geo)physical data files*   Earth scientists have been working on data files since the late 1980s, and were among the first to make them available to the public at large. It was to provide base maps for these global physical data files that the Digital Chart of the World (see below) was first conceived.
- *Environmental data files*   The Global Change Encyclopedia is one of the examples of collections of base data that can be used for environmental purposes. Other examples are the Crop Growth Monitoring System for Europe, based on soil and climate data, which will produce agricultural forecasts with specific climatological input. The output of these programs can often be imported in statistical programs, in order to compare the forecasts with current production statistics.

Before they are entered into the GIS, data from these sources are stored in different ways, some in the form of paper maps, some in the form of files: commercially produced files well protected by copyright (such as boundary files) and published on diskette; remote sensing files sold on tape; socio-economic, environmental or (geo)physical files distributed on CD-ROM; files on oceanographic surveys, photogrammetrically plotted data, or GPS data that are produced only once and are therefore much more valuable. Nowadays, files with spatial data can also be acquired through Internet or the World Wide Web.

It is in relation to the various types of storage or procuring of spatial data that one discerns between analogue maps and other types of virtual maps (after Moellering, 1983). Virtual maps of the first type can be seen but not touched: these are the maps made visible on monitor screens. Virtual maps that are not visible but are tangible are the ones that can be procured on diskette, tape or CD-ROM. These are dubbed virtual maps type 2. The third type of virtual maps are neither visible nor tangible and can be procured over the World Wide Web, such as the maps from the *National Atlas of Canada* (see Chapter 10). As soon as one has queried them from the World Wide Web, they can be viewed on a monitor screen (type 1), or printed as analogue maps (Figure 2.2).

2.2   Vector file characteristics

Though actual file structure will be dealt with in Chapter 4, Sections 2.2 and 2.3 will focus on other aspects of commercially available files used as base maps (boundary files) for GIS maps. Before doing so, vector files and raster files have to be differentiated between. In vector files, lines or boundaries between areas are defined by series of point locations and their

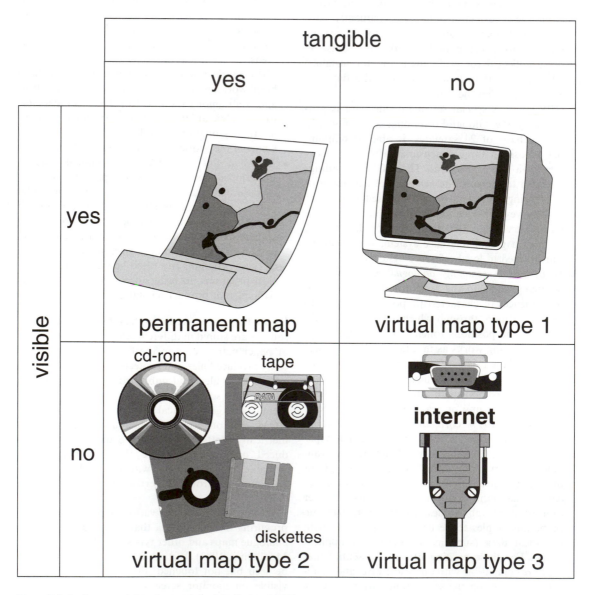

Figure 2.2 Analogue and virtual maps (after Moellering, 1983)

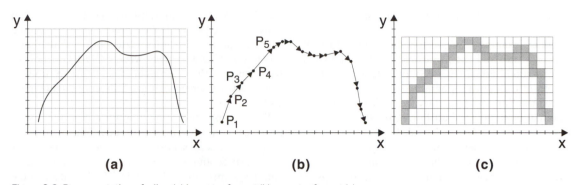

Figure 2.3 Representation of a line (a) in vector format (b) or raster format (c)

connecting links; in raster files, boundaries, or any other relevant background information, are defined as strings of picture elements (pixels) in regular grids, that have been activated with specific values (Figure 2.3).

To use these vector files as geographical reference frames for the spatial information one wants to visualize on a monitor screen, one has to load them into one's computer. When doing so, one has to take into account the following aspects: resolution (the relation between the area as represented by a pixel and the same area in reality; see Section 2.3), digital scale and the possibility of separating the files into different categories.

Vector files, to be used as boundaries, are produced by digitizing maps (see Chapter 4). This procedure takes place on maps of a specific scale. Of course, once the information is digitally available (through digitizing), scale seems to become less important an issue, as it will be possible to zoom in or out and thus change the scale at will. But whenever the scale is increased beyond the original scale, the ensuing image will look poor and coarse, and cannot be used any more for precise referencing. This is because the original map that was digitized will have been subjected to generalization (see Section 5.4). Apart from decreasing the scale, the projection of the map can be changed easily; as soon as the proper transformation formula from one projection to another has been determined, the digitized data can be displayed in any other projection system.

The Digital Chart of the World (a data file containing all the linear elements – coastlines, rivers, contour lines, state boundaries, major roads, and railways and cities) has been produced from Operational Navigation Charts (ONC) at the scale

1:1 million. These ONC charts are ultimately based in most cases on topographic maps 1:50 000, produced from aerial photographs. Both in producing the 1:50 000 maps from the photographs and in producing the 1:1 million charts from the 1:50 000 maps, generalization has been applied with its ensuing simplification, exaggeration, displacement and selection (see Chapter 5). Objects that would still be retained at a scale of 1:100 000 or 1:250 000 are omitted on a scale of 1:1 million. So when the Digital Chart of the World is zoomed in on, and enlarged to a scale of 1:100 000, it would not show all the objects one would expect at this scale.

It would be otherwise if large objects and minor objects could be stored in different layers or files and could be activated whenever a specific threshold scale value was passed. For example, zooming in beyond the scale 1:250 000, minor rivers could be activated and be made visible, etc. So it is important that (boundary) vector files have their objects stored in at least as many layers as there are object categories, and from which a selection can be made for display. An example of such a subdivision of objects in categories would be:

- hydrography: (a) major rivers and lakes; (b) minor rivers and lakes;
- territorial boundaries: (a) national boundaries; (b) state/provincial boundaries; (c) county boundaries;
- coastlines: (a) coastlines and major islands; (b) minor islands;
- administrative names: (a) names of countries; (b) names of states/provinces; (c) names of counties.

The various administrative areas should be supplied with the codes assigned to them, for the purpose of being able to match them to the statistical files (see Section 2.1).

Other examples of vector files are CERCO's Eurostat files and the Electronic Chart Display and Information System (ECDIS). In 1995, CERCO, the Union of European topographic surveys, jointly produced the Eurostat boundary file, which contains the national boundaries of all European countries members of CERCO (Turkey included), and the first, second and third order administrative boundaries. For France, for example, this means boundaries for regions, départements and cantons (groupings of municipalities). The names and codes of these administrative areas have been added as well.

Member states of the International Hydrographic Organization are now developing regional databases for ECDIS (using Digital Data Transfer Standard (S-57), with an common object code and exchange format (DX-90)). Though preference is given to the production of regional databases covering all the routes used by international shipping, this will take time. Limited areas may have data by 1996, but such data will not be available world-wide before the year 2000. In the meantime, hydrographic chart raster files like the one developed in the United Kingdom will be used.

2.3 Raster file characteristics

In a scanner as well as in a fax machine, optical records of the existence of specific dots, dashes or patches in black and white or colour at specific positions are transformed into files with information on positions with attributes. As the registration of these dots, dashes and patches takes place along regular parallel scanning paths, in incremental temporal steps, the output of these files consist of regular grids, built up from picture elements (pixels), each representing specific attributes (Figure 2.3c). The higher the resolution of the scanning device (and its price), the smaller the pixels will be, and more of the detail of the original image will be rendered.

To revert to the case at the end of Section 2.2, existing hydrographic charts are now being scanned and put on CD-ROMs, to provide charts in a form that can already be used in electronic chart display systems, in order to bridge the gap before they will be made available in vector format. The first to do so was the British Hydrographic Office, with its ARCS

package. ARCS (Admiralty Raster Chart Service) consists of CD-ROM files produced by scanning existing charts, with an update option that allows one to combine data from weekly CD-ROM notices to mariners with the original raster chart to produce new charts on screen that are fully corrected (Figure 2.4).

When displayed on the monitor screen the resulting charts can be overlaid with radar images, and can visualize the ship's position as well as data on its course, speed and planned track. The electronic navigation systems using ARCS CD-ROMs will allow the user to select the area to be viewed, to zoom in or out, and to add or omit a number of additional data layers. When the information contained on a paper map is available on different films that can be scanned separately, the raster file can consist of different layers (in this case with contour lines, names, buoys, bathymetric colours, etc.) which can be activated either separately or together.

Apart from updating, the raster image itself cannot be queried or otherwise electronically analysed. It forms a backdrop picture against which spatial processes can be visualized. Important aspects of these backdrop raster files are the same as for the vector files: the scales at which they have been scanned and the existence of information in different layers.

Another example of such a raster file is the raster version of the topographic map of the Netherlands at a scale of 1:25 000. This 'Top25raster' file is sold, per map sheet, on diskette, and contains all the information of the paper topographic map. With products like this the resolution is important: scanners (see Section 2.4) register with a specific number of dots per inch, indicating the smallest areal units for which information is being determined. A resolution of 250 dpi (dots per inch), for example, refers to the fact that areas with a size of 0.01 mm^2 are represented independently (Figure 2.5). As equally important as the geometrical resolution is the radiometric resolution. This refers to the number of colours that can be differentiated by the scanner, and also refers to its display capabilities.

In contrast with vector files, transforming raster files to other projections is very difficult. Coordinate systems with which the images are overlaid cannot easily be changed for other systems. Additionally, the pixel (picture element) is the basic unit of the image structure and it might not be relevant as a reference unit for the theme mapped. A pixel is a representation of the smallest area for which electromagnetic radiation is collected individu-

Figure 2.4 Detail from a raster image generated by ARCS (Admiralty Raster Chart Service) at twice the scanning scale (Courtesy of Hydrographic Office)

ally. The larger these areas, the less information there will be to store and the easier the resultant files will be to handle. On the other hand, the smaller these areas on the earth's surface, the more accurately the resulting raster image will model the original data.

Satellite imagery is also made available as raster files. Resolution here is not a function of the scanner measurement device's size but of the sensor's field of view: by collecting the radiation within a specific angle and from a specific distance from the earth, the sensor registers and measures radiation from a

specific patch on the earth's surface. The size of this patch, the dominant radiation of which is being registered, will determine the usefulness of the imagery for constituting a model of the objects whose radiation values are being registered. These element areas on earth, represented as pixels on the satellite imagery, have decreased in size since the first civil satellites began sending their information to the earth in the 1970s: Landsat first registered 80×80 m squares, and later (Landsat Thematic Mapper) 30×30 m squares; SPOT satellites had a resolution of

analog magnification

map detail

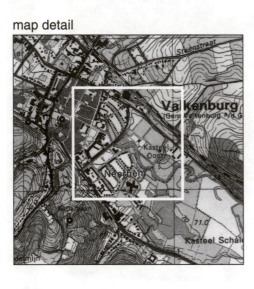

digital magnification

Figure 2.5 Tenfold enlargement of photographs taken from TOP25Raster (the raster-based screen representation of the topographic map of the Netherlands at scale 1:25 000) (left) below a tenfold enlargement of the analogue version of this map (Courtesy Topografische Dienst Nederland)

20×20 m and even 10×10 m for its 'panchromatic' applications. Current plans envisage sending up a SPOT satellite with 5×5 m resolution before AD 2000.

When one compares raster files to vector files, the latter have a greater overall resolution. This is because in the raster technology the input device provides the information divided into discrete pieces ('pixels', from 'picture elements') with a finite size. When the pixels are large, much information is lost. On the other hand, this grid structure of satellite and scanned images allows for analytical operations that are much easier and much less time-consuming than is valid for vector images. It also provides for rela-

tively easy combination with other files, as well as for a multitude of image-processing possibilities. As raster technology is also behind the monitor screens, vector images are simulated in reality on these screens, being built-up from activated raster cells.

2.4 Deriving data from existing maps

The technical procedures for deriving data from existing maps will be described in Chapter 4. In this section the requirements existing maps have to answer to, in order to be suitable sources for digital files, will be indicated. It is especially documentary aspects that are important here. Such a plethora of paper maps has been produced that finding the proper one can sometimes be extremely difficult. Furthermore, the organization and aims will be covered, while the principles of the hardware used will be described. The data to be derived from existing maps are complex in the sense that they have both locations and other attributes. So as a first requirement these locations and attributes should be unambiguous on the source documents: the definitions used or implied in the map legend should be clear and consistent, and the period of time over which the data were gathered should be mentioned (in order to be copied and stored somewhere in the digital files as well). The data quality aspect is as important, and will be covered in Section 2.5.

2.4.1   Finding the proper map: documentation

Finding the proper map will depend on the selection of a map of the proper area, the proper theme and the proper time period in which the data have been collected. Most map libraries will contain map descriptions in their catalogues that refer to the

area and theme of a map, and also have information on the map's date of publication. They will rarely have information in the catalogue files on the time period in which the data for the maps was collected, i.e. the period for which the map is valid. For topographical maps this is usually around two years before the map was published; for thematic maps it might be longer. A good indication of the timeframe of international map products, be they in paper or in digital form, is provided by the Digital Chart of the World (DCW) released in 1992. As this file has been digitized from 1:1 million ONC charts, Figure 2.6 shows the survey timeframe of the map sheets for the world, indicating those areas in which the digital information was based on maps published prior to 1970 and those areas based on maps published after 1980. The DCW fits on four CD-ROMs, containing more than 1.6 gigabytes of data. Its contents are split over several layers, such as international boundaries and coastlines, land use, roads, railroads and hydrography. Each layer is split over several sub-layers. For instance, railroads are split into single and double track lines. Access to the data is available through a place-names index, index maps, or by giving a position in longitude and latitude. These data are topologically structured in the Vector Product Format, and output to those GIS packages that can handle

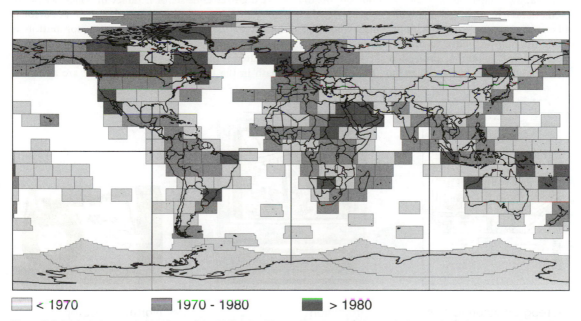

☐ < 1970   ▨ 1970 - 1980   ■ > 1980

Figure 2.6 Period of survey index sheet of the ONC 1:1 million map series used for producing the files of the Digital Chart of the World

VPF is part of the international exchange standard DIGEST.

2.4.2 Preparation

Finding out about the way in which locations have been indicated on the paper map is essential, as this will be a guide for the manner in which the information to be digitized or scanned will have to be transformed in order to fit in one's GIS. When preparing an existing map for either digitizing or scanning, the method in which the data are spatially referenced should be taken into account. This georeferencing (providing a spatial address) may have been done by using geographical coordinates, grid coordinates or no coordinates at all. In the last case, the map cannot be further fitted into or integrated with existing files, unless by rubber sheeting (see Section 5.3). When using geographical coordinates the earth is regarded as a sphere; when using grid coordinates, the area represented is regarded as a flat plane (see Section 5.1). The UTM grid coordinate system combines both geographical and grid coordinates in a sense, as it consists of a number of planes that together cover the earth (see Section 5.2).

If, for example, a file digitized at scale 1:250 000 of northwest Europe has to be extended or updated, topographic maps on a scale of 1:50 000 of the Netherlands have to be digitized. This means that all coordinates implied in the file produced when digitizing the map will have to be transformed from the stereographic azimuthal projection the maps of the Netherlands are rendered in to the Universal Transverse Mercator projection of the 1:250 000 file. Digital files are scale-less in principle, but their resolution is still governed by the scale at which the data were digitized. Now, in order not to have discrepancies in the resolution, for the data from these maps of the Netherlands to fit in the new 1:250 000 Northwest Europe file, they have to be generalized as well (Figure 2.7).

Part of the preparation for digitizing complex analogue maps entails highlighting the lines that have to be digitized with a marker, and entering manually the codes that will be applied to the lines and to the areas they separate on the sheets. The direction in which the lines will be digitized will be determined and indicated as well (e.g. clockwise, from top to bottom, etc.) so as to result in a standardized procedure that is logical and which can be continued by others if need be. A procedure should also be prepared for indicating which lines and points have been digitized already.

2.4.3 Digitizing

The main means of converting analogue data into digital data are manual digitizers and scanners. For digitizers, manual input is still part of the procedure. Scanning is done automatically.

A digitizer consists of a tablet in which wires are embedded, located along Cartesian axes. A cursor is linked to the tablet and, when a cursor button is pushed somewhere on the tablet, the electrical charge generated is picked up by those wires directly underneath in the tablet, and the wires that are activated

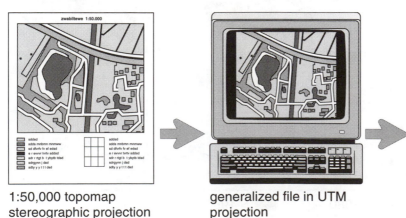

1:50,000 topomap
stereographic projection

generalized file in UTM
projection

file 1:250,000
Northwest Europe

Figure 2.7 Conversion of an analogue map into a digital file

then provide their specific codes as X and Y coordinates. Additional buttons on the cursor allow one to join attribute information to the locational information. For entering topological information (information about logical relationships between spatial objects, such as relative positions, adjacency and connectivity), like stating whether a node belongs to a specific line, or is the beginning or end point of a line or is located somewhere in-between, specific areas on the digitizing tablet may be activated and used as legend boxes. By pressing certain areas on the screen by hand, or by clicking on the area of the screen using a mouse, the link between the digitized location and specific attribute information is forged and entered into the computer memory. Such areas on the digitizing tablet are collectively called menus.

The accuracy of these digitizing tablets is a function of the distance between the wires in the tablet; their functionality is also determined by their size, which can vary from A4 to 1.5 × 2 m (Figure 2.8).

The main objective during this process of converting spatial information from analogue into computer-readable form is to preserve the relationships that were visualized on the map. This means, for example, that

- existing links between points should be retained,
- parallel lines should remain parallel,

- relative locations should be preserved,
- absolute locations (as expressed in coordinates) should be preserved,
- adjacency should be preserved, and
- lines that merely touch each other should not intersect each other, etc.

For 'professional' digitizing of large series of maps, configurations are used which are ergonomically adapted to the working attitudes, allowing for a minimum of tiresome bending over, while the digitized lines and points will be visualized immediately on-screen, so that the operator can check on-line whether he or she is following the marked information properly. This checking is very important because it will also decide on the usefulness of the resultant files, as the potential for mistakes is very high. Current digitizing software also provides for on-line control of the digitizing input. Examples of these possible mistakes during digitizing are given in Figure 2.9.

An inadvertent move of the reference point of the digitizer will result in an affine transformation of the coordinates. Digitizing the same line twice will almost inevitably lead to sliver polygons (or slivers, i.e. narrow meaningless areas). Sudden movements of the cursor not consciously registered by the operator during digitizing will lead to spikes, as might the

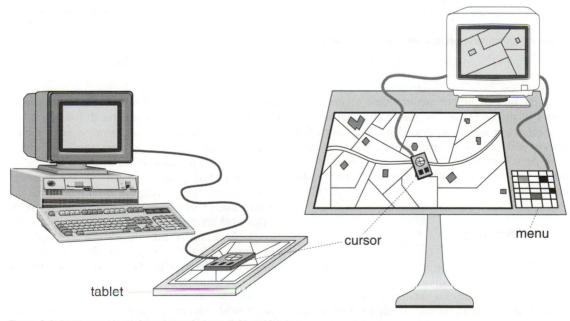

Figure 2.8 Digitizer tablet (left) and stand-alone digitizer (right)

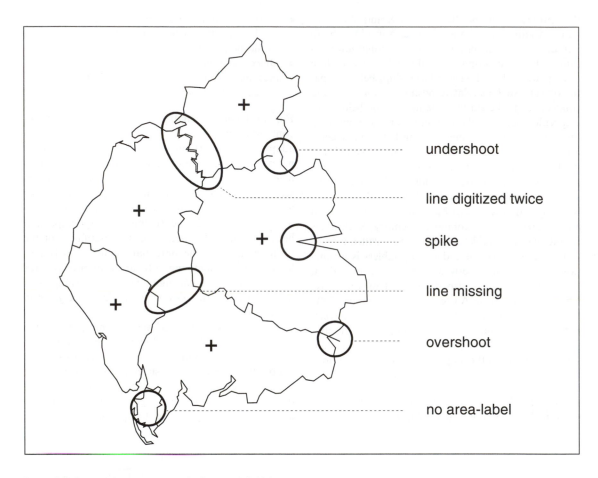

Figure 2.9 Potential errors as a result of manual digitizing

addition or deletion of points. Line segments or nodes might be missed as well. Lines that should connect might not do so after digitizing. Labels for lines or nodes or for the areas the lines separate might have been forgotten, and links might overshoot or undershoot the nodes where they are supposed to end.

2.4.4 Scanning

Mistakes like these can be avoided by using scanners (Figure 2.10). The operating principle of high-resolution scanners is that of a rotating drum to which a paper map has been attached. While rotating, a sensor will scan the map in narrow (e.g. 0.1 mm wide) contiguous bands, and will register the light intensity (or colour value) of small squares (0.1 mm^2). These light measurements are then transformed to digital values, and can be digitally represented on screen, thus reconstituting the scanned original. By splicing the signal picked up by the sensor over measurement devices susceptible to red, green or blue light, the original colours can be reconstituted (Figure 2.10b).

The working principle of low- to medium-resolution table scanners used in desktop cartographic environments is that of registering the characteristics of an image on a page, put face-down on an A4 or A3 tablet, line by line. All data points or pixels on a line are registered sequentially before moving to the next line. It subdivides images into discrete data points of which the sensor measures their light value or colour

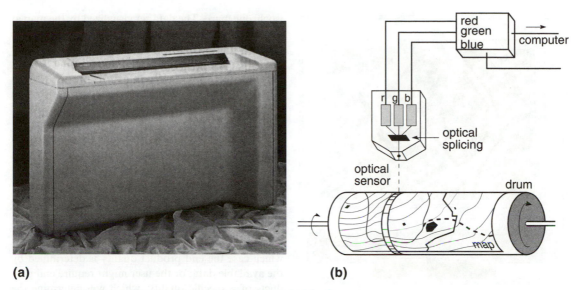

(a)   **(b)**

Figure 2.10 (a) A scanner (Courtesy Intergraph); (b) working principle of a scanner

value. In this way an analogue map is turned into a description consisting of spatial addresses (grid cells) and their characteristics (light values).

This is the easy part of scanning. After scanning, there is still the operation to make sense of the scanned data, and to change it back into a vector file, if necessary (see also Chapter 4). If possible, the printed maps themselves are not used for scanning; rather, the original colour separates prepared for the map's reproduction are used. On these, there is already a separation of functions which should then be further edited. On scans from the separates to be printed in blue one should indicate whether lines refer to rivers, coastlines or lake shores; codes can be added to the rivers that should indicate their importance; this is relevant when the scanned image should subsequently be generalized and represented on a smaller scale.

At present (1995), heads-up digitizing is considered to be the best procedure. It is a combination of scanning and digitizing. For this procedure the map as a whole is scanned, and then displayed and enlarged on a monitor screen. The lines to be scanned can be highlighted, and the operator will then follow/trace and enter with the cursor the relevant lines or point locations on the screen. The parts of the lines entered will be displayed in a different colour; the image can switch between that of the scanned map, partly digitized, and that of the digitized information only,

allowing the operator to better check the consistency of work done (Figure 2.11).

2.5   Control and accuracy

The advent of GIS has accelerated and simplified the process of information extraction and communication. Combining or even integrating various data sets was made possible on a large scale. The ease with which operations can be effectuated provides a danger as well, as technical possibilities will also allow for irrelevant or inconsistent data integration. On the other hand, the new storage potential provided by new PCs will allow one to store the original data and not the derived or aggregated data. Figure 2.12 provides an example of the difference betwen using original and aggregated data. Here, population distribution data (visualized as a dot map, showing locations of specific numbers of inhabitants) and population density data (visualized as a choropleth map, showing densities for enumeration areas) are both used as a starting point for an analysis of the average distance the inhabitants of a region have to walk in order to reach specific municipal facilities. Therefore, both data sets are combined with data on the average travelling time (here expressed as an

Figure 2.11 Heads-up-digitizing

isochrone map). The data set visualized by the map at the lower left gives a distinctly different image to the image that results from starting with the data set visualized at the upper left. If decisions about siting new facilities are based on either of these maps, it is important that the decision-maker is given enough information to be able to ascertain the quality of both maps.

The end product of every type of date integration will inevitably have a certain degree of uncertainty, because of mistakes in the original data and because of the data processing. The result will only be fit for use when a certain level of reliability is reached. That is why the user needs information on the quality of the original data and of the various processing steps.

The user might choose to depend on the data, in which case the end product quality is determined by the available data; or the user might require end products of a specific quality, which will determine the data quality of the original data to be collected and of the processing methods to be followed

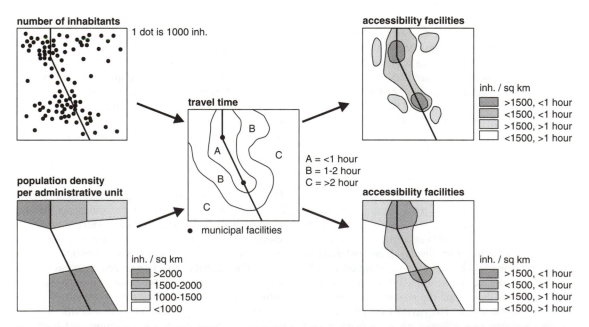

Figure 2.12 Results of data integration. If the aggregated data set is used (bottom), the densities of the population able to reach the facilities within a specific time will be visualized differently as compared to the map based on distribution data (after Hootsmans and van der Wel, 1992)

Traditionally in cartography the quality of the original data was expressed in the form of reliability diagrams. These consisted of statements that described the manner in which the data for specific parts of the area mapped had been collected, and did not differentiate within these parts (Figure 2.14b). Thematic equivalents of these topographic procedures were statements such as 'soil samples have been taken for every hectare' or 'map unit boundaries should be accurate within $\pm$ 100 m'. But for correct decision support, data quality information is not only needed on an overall level, local/regional deviations from the overall accuracy have to be provided as well. General accuracy information will help the user to place the data sets in order of their importance regarding decision-making, while local anomalies in accuracy will be used to restrict the decision-making process to specific areas.

In the literature (DCDSTF, 1988) the following aspects of data quality or accuracy are discerned between: lineage, positional accuracy, attribute accuracy, logical consistency and completeness. Lineage will indicate when the data were collected, and what kinds of processing they were subjected to. Logical consistency will only refer to data sets as a whole, while the other aspects might deviate locally. This will allow one to visualize their spatial variation in map form.

In satellite data, various land cover categories or classes can be differentiated between on the basis of the spectral characteristics of the radiation measured from their pixels and from field checking them in the terrain (training). On the basis of a number of trained categories, for every element (pixel) of a satellite data set a probability vector can be determined, defining the probability of this pixel being assigned to each of the categories discerned. The category with the highest probability value will be selected firnally for representation.

Probability values vary between 0 and 1. The category with the highest probability value is called maximum likelihood, the category with the second highest probability value is called second likelihood, etc. When assigning a category to a pixel, it will make quite a difference whether the maximum likelihood is 0.9 and the second one 0.1 or whether the maximum likelihood is 0.3 and the second likelihood 0.28. In the latter case the certainty of assigning this pixel to this category is much lower, which is why uncertainty information can be added, in order to evaluate the resulting patterns.

Uncertainty information can be visualized in one of the following map forms:

- probability images, such as maps of the maximum likelihood, maps of the second likelihood, etc.,
- maps of the difference between maximum and second likelihood, or
- maps of likelihoods for each category discerned.

Apart from being visualized, it can also be rendered by sound: shrill sounds generated when moving a cursor over a less certain area might, for instance, indicate low data quality. A measure for uncertainty or ambiguity could be the confusion index, in which the maximum likelihood is compared to the second likelihood.

$$\text{Confusion} = 1 - (m_{\max} - m_2)$$

Small differences between $m_{\max}$ and m_2 will lead to high values for confusion, which can also vary between 0 and 1.

If classes cannot be sharply defined, it is conceivable that objects have not only a specific certainty (possibility) to belong to one category, but also another certainty (possibility) to belong to other classes/categories as well. This is expressed by the fuzziness index, which compares possibilities (these can vary from 0 (not member of a category) to 1 (member of a category)) to Boolean membership values, which can only be 0 or 1, as in Boolean logic the objects are defined in an either/or way, leading to crisp boundaries between objects. So if, because of overlap between category or class definitions, a location may be assigned to different categories, in a situation where it is not imperative that it be assigned to one class only, possibility vectors can be determined.

In Figure 2.13 maps are shown in which only the uncertainty zone of the class boundary is visualized for different certainty threshold values as an alternative for conventional 'crisp' boundaries.

Next to probability and possibility factors, certainty factors are assigned on a more arbitrary basis, i.e. by experts who estimate the validity of data, and express these values also on a scale from 0 to 1. All the different methods of representation can be expressed as shown in Figure 2.14. It is the ideal that it would be possible to ask for information as contained in Figure 2.14, to toggle it with the visualized data themselves, in order to allow for proper decision-making on the basis of the data.

uncertainty threshold = 0
Arrows point at class boundary locations, this visualization suggests unconditional certainty of boundary locations.

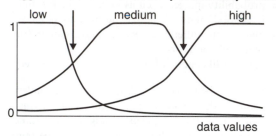

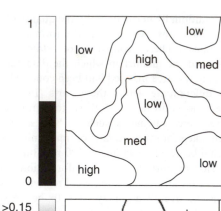

uncertainty threshold = 0.15
Fuzzy region in graph indicates boundary transition with a certainty of less than 0.15 (corresponding grey tones in graph and map).

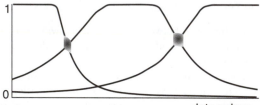

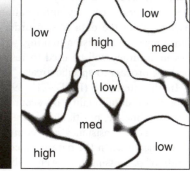

uncertainty threshold = 0.30
Fuzzy region in graph indicates boundary transition with a certainty of less than 0.30 (corresponding grey tones in graph and map).

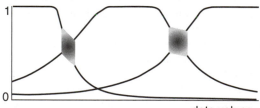

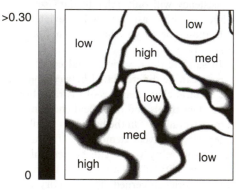

uncertainty threshold = 0.50
Fuzzy region in graph indicates boundary transition with a certainty of less than 0.50 (corresponding grey tones in graph and map).

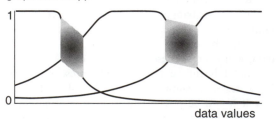

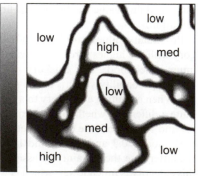

Figure 2.13 Uncertainty zones of boundary values, in which the overlap between class definitions is expressed as grey tones in the map (after Hootsmans and van der Wel, 1993)

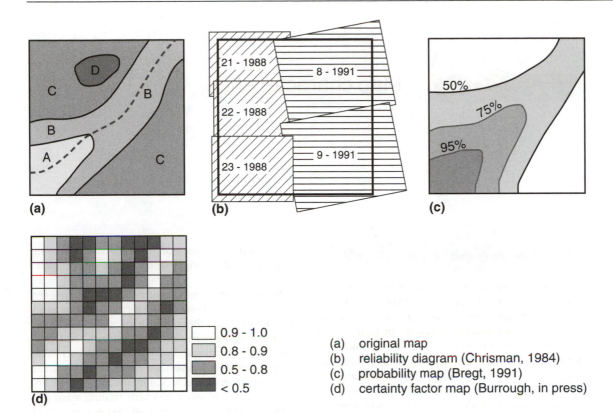

Figure 2.14 Examples of the visualization of meta-information (from Hootsmans and van der Wel, 1992)

(a) original map
(b) reliability diagram (Chrisman, 1984)
(c) probability map (Bregt, 1991)
(d) certainty factor map (Burrough, in press)

0.9 - 1.0
0.8 - 0.9
0.5 - 0.8
< 0.5

Product references

ARCS (Admiralty Raster Chart Service) (1994) Hydrographic Office, Taunton.

CD-ROM Atlas de France (1991) GIP/RECLUS, Montpellier.

Crop Growth Monitoring System (1993) Produced for Joint Research Centre, Commission of the European Communities, by the Winand Staring Centre for Agricultural Research, Wageningen, Netherlands.

Digital Chart of the World (1992) DMA/ESRI.

Geoscope Global Change Encyclopedia (1993) Canada Centre for Remote Sensing, Ottawa.

National Atlas of Canada: E.M. Siekierska (1993) From the Electronic Atlas System to the electronic atlas products. Electronic Atlas of Canada from the beginning to the end. In I. Klinghammer *et al.* Proceedings of the seminar on electronic atlases, held in Visegrad 1993. Budapest: Eötvös Lorand University, pp. 103–111.

ONC – Operational Navigation Chart. Aeronautical Chart and Information Center, St Louis, USA.

CHAPTER 3

Map characteristics

3.1 Maps are unique

It is difficult to get an overview of an area in any way other than by consulting a map. A map places spatial data, i.e. data about objects or phenomena for which the location on the earth is known, in their correct relationship to one another. A map can be considered as a spatial information system that gives the answers to many questions concerning the area depicted: the distances between points, the positions of points in respect of each other, the size of areas, and the nature of distribution patterns. The answers can be read off directly from the map image most of the time, without the need for a keyboard or the loading up of some files.

Theoretically, geographic information systems would be able to work out solutions to problems set to them without maps, just on the basis of spatial information in digital form which may have been collected as such (though in most cases it would have been digitized from a map). But performing such a task would be questionable in practice, as without maps one would hardly be able to formulate the relevant spatial problems that can be solved using a computer.

The term 'map' is used in many areas of science as a synonym for a model of what it represents, a model which enables one to perceive the structure of the phenomenon represented. Thus mapping is more than just rendering, it is also getting to know the phenomenon that is to be mapped. By 'cartographic method' one understands the method of representing a phenomenon or an area in such a way that its spatial strucure will be visualized and this will usually take some experimenting. When representing spatial information in map form one has to limit oneself, on account of the available space, to the essentials, and amongst which is the information's structure.

One of the most important moments for cartography was when the first satellite pictures became available. This created the opportunity to check whether the mapping activities of past centuries, and especially the generalization from large-scale detailed maps to small-scale overview maps of larger areas, had been done correctly. Comparison of those first Apollo satellite photographs with existing maps generally showed a great deal of agreement, which proved that the techniques applied had been appropriate.

A second important moment for cartography was the introduction of the computer. Initially (in the period 1960–1980) the computer was used to automate existing mapping tasks, such as calculation of projections and the plotting of the grid or graticule on the map. It proved to be feasible to map an area according to different projections, based on the combination of the same digital file with different transformation parameters. Gradually, cartographers also came to realize the potential for analysis inherent in the digital (digitized) data which computers offered. It then became clear that with the aid of the computer one would be able to do calculations with the digitized map data, and that one could have the computer determine distances, areas and volumes much more precisely than one could do oneself using paper maps.

As soon as a link was made between these cartographic (boundary) files and data files it also became possible to evoke numbers of inhabitants, average income data or agrarian production figures and combine them digitally with the cartographic files in map form; the same could be done for the relation between certain socio-economic and physical phenomena and topography. This has developed into cartographic information systems, which operate similarly to the geographic information systems defined in Chapter 1, but are geared more to visualization than to analytical functions. We must keep reminding ourselves that it is the capacity for abstraction of the map that allows one to perceive spatial connections, patterns or structures. It does not matter whether the map is visualized on paper or on a monitor screen.

Maps are nowadays regarded as a form of scientific visualization (Figure 3.1), and maps indeed existed already before visualization developed into a distinct field. The objective of visualization is to analyse information about relationships graphically, whereas cartography aims at conveying spatial relationships. Visualization consists of graphics (with which symbols and lines are indicated) and geometry which refers to their relative positions. In cartography these relative positions are usually defined on the basis of a spatial grid – cartesian or geographical – which relates these locations to real positions on the earth's surface. The emphasis in scientific visualization (McCormick *et al.*, 1987) is more on its analytical power ('explorative analysis', see Section 9.5) than on its communicative aspects: it is primarily directed at discovery and understanding. In cartography, emphasis can lie equally on analysis and communication.

An example of the difference between the analytical and communicative aspects of cartography is shown in Figure 3.2. Here, on the right, the distribution of the precipitation classes has been visualized, and this has been done correctly, as it is possible to see in one glance the distribution of every class in relation to all other classes. But getting an idea of the distribution of the various precipitation classes is not the same as conveying to the reader a proper idea of the distribution of the phenomenon 'precipitation' itself. If it is the communication objective to show the effect of a particular thunderstorm over the Netherlands, then one would want to perceive the spatial trends, the increases and decreases in its effects over the area, and that is something different from providing an idea of the location of the various precipitation classes. The wrong graphic variables have been chosen to answer that objective in Figure 3.2(b). It is in Figure 3.2(a) that the correct graphic variables have been selected, so that the increase in grey values is proportional to the increase in precipitation values, which makes the map fit for the communication objectives because it portrays the phenomenon as a whole. Figure 3.2(b) is only fit for analysis, independent of the phenomenon. This would be acceptable in a data exploration environment, where the user is in command of the display time and would be able to adjust the class boundaries at will.

One of the most important aspects of exploratory analysis in cartography is the change of perspective that is provided through transformations. In cartography, these transformations are effectuated by deviating from traditional map projections, by deviating from Euclidean geometry by representing other than geographic distances and areas, by deviating from real life by exaggerating the values or measurement data to be represented (as happens, for example, in three-dimensional models where the vertical scale might be ten times the horizontal scale), by separating local from regional trends, so as to make the latter stand out, or by experimenting with class boundaries. Other ways of changing perspective are by deviating from traditional geographic map frames (through scrolling in electronic maps), by changing the orientation of maps (by not having the north at the top) as well as their time frames (by monitoring) so that the effects of a random survey activity or of a random data gathering activity are offset.

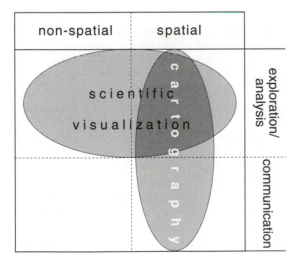

Figure 3.1 Relationship between scientific visualization and cartography

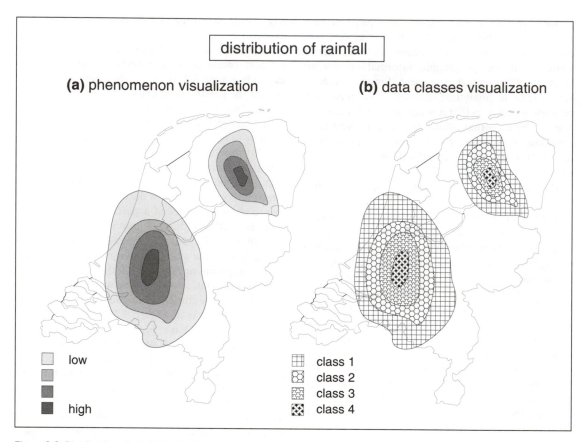

Figure 3.2 Distribution of rainfall in the Netherlands: example of map designs geared to either (a) data communication, or (b) data analysis objectives

3.2   Definitions of cartography

The meaning of the term 'cartography' has changed fundamentally since 1960. Before this time cartography was generally defined as 'manufacturing maps'. The change in the definition is due to two factors: first, the fact that the subject has moved into the field of communication sciences, and secondly, the advent of the computer. Cartography nowadays is seen as '*the conveying of spatial information by means of maps.*' This results in the view that not only the manufacturing of maps, but also their use is regarded as belonging to the field of cartography. And it is indeed evident that only by investigating the use of maps and the processing of the mapped information by their users is it possible to check whether the information in the maps was represented in the best possible way.

The unsatisfactory aspect in the definition above of 'the conveying of spatial data by means of maps', is that the concept 'map' has not yet been defined. The elements which belong in a definition of maps are spatial information, graphic representation, scale and symbols. A possible definition of a map runs as follows: *a graphic model of the spatial aspects of reality*. According to French cartographers, the map is *a conventional image, mostly on a plane, of concrete or abstract phenomena which can be located in space*. By 'conventional', it is meant that one works with conventions, such as the fact that the sea is represented in blue, that the north is at the top of the map, or that some graded series of circles denotes settlements with increasing population numbers. By 'image', the graphic character of a map is stressed. But not all maps are printed on a sheet of paper: relief models and globes are also considered to be maps. It is, of course, also possible to map phenomena that are not physically tangible, such as political preferences or language-borders. And it is obvious that it must be possible to locate the phenomena in space.

Under the influence of the rise of the computer and geographic information systems in the field of mapping, new definitions of cartography have gradually emerged: 'the information transfer that is centered about a spatial data base which can be considered in itself a multifaceted model of geographic reality. Such a spatial data base then serves as the central core of an entire sequence of cartographic processes, receiving various data inputs and dispersing various types of information products' (Guptill and Starr, 1984).

Taylor (1991) defines cartography as 'the organisation, presentation, communication and utilisation of geo-information in graphic, digital or tactile form. It can include all stages from data preparation to end use in the creation of maps and related spatial information products'. As this still requires 'map' to be defined, Board's (1990) map definition is quoted: 'a representation or abstraction of geographic reality. A

tool for presenting geographic information in a way that is visual, digital or tactile.'

The usefulness of a map not only depends on its contents, but also on its scale. The map scale is the ratio between a distance on a map and the corresponding distance in the terrain. There are several possibilities to indicate the map scale, as can be seen in Figure 3.3. Next to a verbal description (such as one inch to the mile), a representative fraction (such as 1:1000) or a graphic representation can be used. The representative fraction in Figure 3.3(a) means that 1 cm on the map corresponds to 1000 cm (or 10 m) on the terrain. The scale bar in Figure 3.3(b) represents a distance of 10 km. When the representative fraction is small, a map is considered to have a large scale. Figure 3.3(a) shows a large-scale map of the city centre of Maastricht, scale 1:1000. The map reveals many details on the level of individual houses. Figure 3.3(b) is a detail from a small-scale map with a scale of 1:500 000. Again, the map displays some data on Maastricht. Here the whole urban area of the city is shown as a small polygon, however.

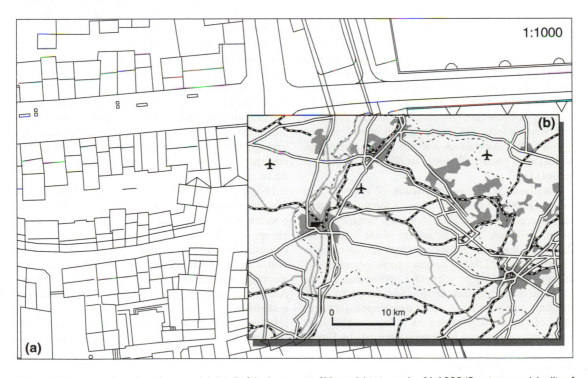

Figure 3.3 Large- and small-scale maps: (a) detail of the base map of Maastricht at a scale of 1:1000 (Courtesy municipality of Maastricht); (b) Maastricht at a scale of 1:500 000

Confusion exists about the concepts large scale and small scale. In every day linguistic usage, small scale is linked with small areas but in mapping this is used in the reverse sense: small scales in cartography are linked with large areas that are represented on a small map area (Figure 3.3b). A large scale in cartography is connected with a small area, with detailed data presented on a relatively large map area (Figure 3.3a). Technically, the linguistic usage in cartography is correct: a large scale is represented as a fraction that has a relatively small figure in the denominator; a smaller scale represents a smaller value, thus a bigger figure in the denominator.

In Figure 3.4 the effect of changes in scale is demonstrated. When scales are changed, generalization becomes necessary, as not all the information from large-scale maps can be incorporated in small-scale maps. In socio-economic thematic maps this generalization usually takes the form of aggregation: small enumeration units are grouped together to form larger units. In Figure 3.4 the election wards/polling districts in the large-scale map are grouped together to form municipalities in the medium-scale map of the Netherlands. This aggregation has a direct effect on the data, in this case the Labour Party vote percentage: as the figures are counted for larger areas, excessive local scores are averaged, and differences become less notable. In the small-scale map these differences are still visible because the economic regions are groupings of municipalities on the basis of common economic characteristics (e.g. agricultural versus industrial regions, with the ensuing political preferences). If the last grouping had been effectuated according to nodal regions (groups of municipalities oriented on the same large towns) then these differences would have been even less obvious.

Traditionally, the main division of maps is into topographic and thematic maps. Topographic maps supply a general image of the earth's surface: roads, rivers, buildings, often the nature of the vegetation, the relief and the names of the various mapped objects (Figure 3.5a). Thematic maps represent the distribution of one particular phenomenon. In order to illustrate this distribution properly every thematic map, as a basis, needs topographic information; often this is provided by a topographic map where minor features have been omitted. In Figure 3.5(b) one can find a detail of a soil-map as an example. Such a map is only functional if one is able to locate (with the help of this topographic base map) where the different soil types can be found.

A thematic map would also emerge if one aspect of the topographic map (such as motorways or windmills) is highlighted, so that the other categories of data on the map are perceived as ground. An example of this is presented in Figure 3.6.

In a digital environment the differentiation between topographic maps and thematic maps is less relevant, as both map types consist of a number of layers: a topographic map would be a combination of separate road and railway layers, a settlement layer, hydrography, a contour-lines layer, a geographical names layer and a land cover layer. Each of these layers would be a thematic map in itself, and a combination of layers, in which each data category had the same visual weight, would be a topographic map. If one category were to be graphically emphasized or highlighted, and the others thereby relegated to the status of ground, then it would again change into a thematic map.

The topographic base of a thematic map can even be far more schematic than an excerpt from a topographic map. In the representation of socio-economic phenomena, the data are gathered for enumeration areas instead of at individual locations (as would be the case for physical phenomena), the boundary files are usually strongly generalized, so as to distract from the map theme as little as possible. An example is shown in Figure 3.7. In Figure 3.7 the base map only consists of national boundaries and coastlines. Such a presentation (the combination of thematic data (Figure 3.7b) plus schematic base map (Figure 3.7a)) leads to a clear map image (Figure 3.7c), which can only then be interpreted and memorized correctly when the map user also recognizes the area. To arrange for this recognition it may be necessary to add the names of the (most important) enumeration units.

People form for themselves a mental model of reality. For example, whilst living in a village one would operate, while traversing the village for one's daily chores, on the basis of a mental construct, which operates like a *virtual map* (Figure 3.8). When a visitor asks for directions, one would consult this mental construct in order to provide an answer. Answering the request could also be done by drawing a sketchmap, a '*mental map*' which is a permanent print-out from our mental construct of reality, designed for answering a specific request for directions from and to random locations. Such a mental construct cannot only be generated from one's contact with reality, but also through consulting a proper, tangible paper map (a '*permanent map*', or

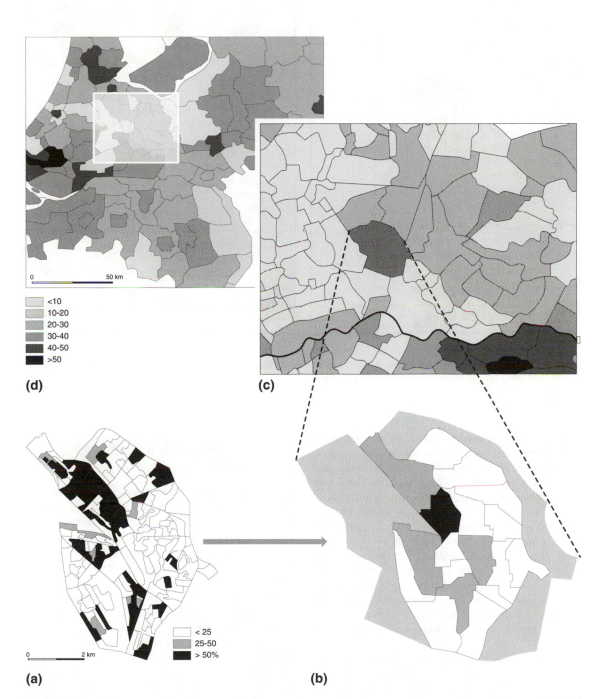

Figure 3.4 Thematic maps and the influence of scale on socio-economic phenomena: the percentage of Labour votes in (a) Utrecht's voting districts; (b) neighbourhoods; (c) municipalities; (d) regions

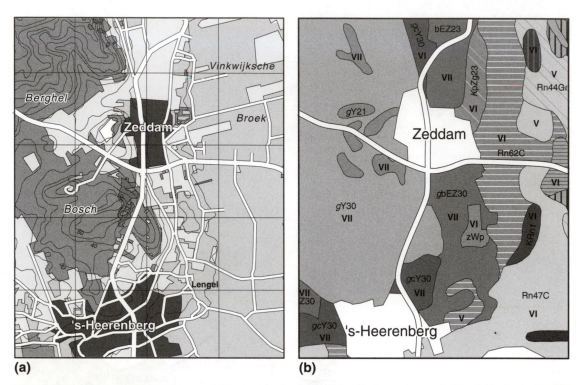

(a)   (b)

Figure 3.5 Topographic and thematic maps: (a) the topography of Montferland in the eastern part of the Netherlands; (b) a soil map of the same area (after Netherlands soil map 1:50 000)

hardcopy map in computertalk). When one is an experienced map user, one would be able to grasp the essential information from a paper map and store this is one's mind. This process can also take place from a map displayed temporarily on a monitor screen (a '*temporal map*', which is visible but not tangible). This temporal map may be generated from a spatial database stored in the computer, from which a specific selection has been made in order to answer specific requirements or objectives. This spatial database, which can also be used to produce other maps, functions as a *virtual map* therefore, in the same way as different sketch maps can be produced from one's mental perception of reality (Figure 2.2).

Maps can be said to show three dimensions of the phenomena represented: the nature or the value of the objects (that is their attributes) and their location. The location is defined using the x- and y-coordinates. It is one of the tasks of cartography to have the representation of the attributes stand out sufficiently on the flat paper or monitor screen surface.

3.3   The cartographic communication process

In Section 3.2, cartography is described as 'the conveying of spatial data by means of maps'. In order to illustrate this, a model of cartographic information-transfer is presented in Figure 3.9.

The starting point of the cartographic communication process is the data or the information (I), usually collected by third parties (e.g. geodesists, photogrammetrists, geographers, statisticians). Cartographers have to study this information, as they will have to get acquainted with it, as well as with the purpose of the information transfer, before they will be able to represent the information correctly in map format. Often the resulting map will not contain every particle of information with which the cartographer was supplied; generalizations or classifications may have been applied in order to present a clearer picture of the phenomenon. The map user or map reader who sets eyes upon the map will derive certain informa-

Figure 3.6 Thematic map created by subduing part of the topography, and by doing so highlighting a specific information category (highways)

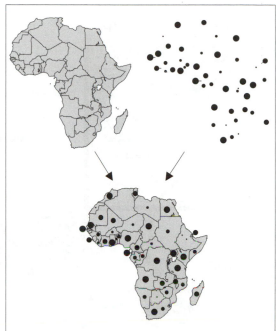

Figure 3.7 Compilation of a thematic map (see also Figure 4.5)

tion from it. The information derived (I') will, however, never completely overlap or coincide with the original information (I) as during the communication process data may have fallen out or been left out on purpose, or because mistakes may have been made. The cartographers may have interpreted the original information incorrectly; and even if they did interpret it in the right way, then they may still have made errors in the process of mapping, when representing the information. The map reader may read-out the data in the wrong way, or may draw the wrong conclusions from the right data. Thus there are ample possibilities for the derived information not to coincide completely with the original information.

Cartography aims for the elimination of these various sources of errors, and for the provision of a correct transfer of the data, by means of such a graphic presentation that the map reader is able to draw the correct conclusions. A feedback of I' to I in Figure 3.9 is necessary, because in this way the cartographer is able to check the effect of the cartographic

products and depending on the evaluation, may adjust the image of the map. During the process of evaluation, moreover, one keeps on learning about the depicted phenomenon: whether the correct structures have been transferred, whether the most relevant characteristics have been selected, and whether the most recent quantitative data have been represented. Therefore the cartographic representation is also a cognitive process: one has to get to the essence of a spatial phenomenon, if it is to be represented adequately.

Of course, in principle maps are a representation of the earth's surface: when viewing a map we visualize the earth's surface, and then may try to match our mental vision with the real world. The overlap referred to earlier should also apply to the real world but this overlap can never be complete: there will always be a time gap between the moment the data were gathered or surveyed and the moment they were made available to the users. In some cases (e.g. in recent military conflicts) there may only be a

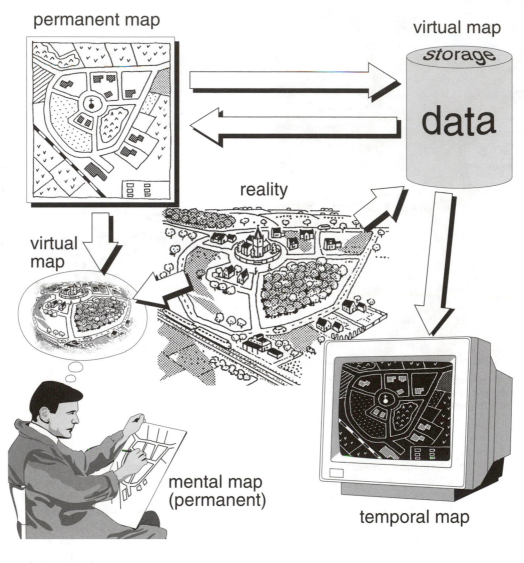

permanent map

virtual map

storage

data

reality

virtual map

mental map (permanent)

temporal map

Figure 3.8 Examples of permanent, virtual, temporal and mental maps

couple of hours between data collection and the same data being made available, but even then the time gap may be crucial as the planes stationed on an airfield or tanks on the ground may have left the area in the meantime.

The cartographic information process, as presented in this book, thus starts with the recording. The traditional, cartographic 'recorder' is the topographer, who surveys the terrain for the sake of the production of the topographic map, and who describes functions, distinguishes road categories, collects geographical names, measures the location of new objects, etc. The topographer's work has already been referred to in Chapter 2, and the results of this work will be discussed in Chapter 5.

Topographers (together with geodesists) supply the x- and y-coordinates of the spatial information; the attribute data (with the exception of altitudinal data) are defined by others: by census-takers who collect information by means of questionnaires, especially of a socio-economic nature, by soil scientists who take samples of the soil, or by traffic enumerators etc.

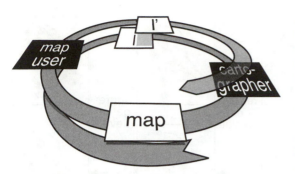

Figure 3.9 Model of the cartographic communication process

Since the 1930s aerial photographs have played a very important role as a data source, while since the 1970s satellite images have been increasingly important also.

From all these sourses, traditional and new, a lot of information is directed at the cartographer, most of it not in the right configuration for cartographic presentation. Thus some kind of (cartographic) processing is necessary first. The elaboration of the survey data, in many cases, takes the form of a classification. Individual characteristics of observations are then replaced by group characeristics. Thereby the quantity of information decreases, and the resulting map becomes better ordered. Other methods of elaboration are the correlation with other data, the expression of the degree of conformity or difference in the distribution of different data sets, or the comparison of absolute enumeration data taken at different times for the same area, in order to be able to show the development in the intervening periods.

The design of maps is largely concerned with making choices: choice of mapping method, and choice of the graphic variables (such as differences in size, value, grain, colour, direction and shape; see Chapter 6) to be used. In Figure 3.10 a number of selection moments are shown, which play a role in the design of maps, as well as the cartographic results of these choices. Here the development of the population of the Netherlands in the period 1970–1980 is concerned. The first choice that has to be made is the one between map and diagram. Should the diagram be selected, then the opportunity to show relations to other spatial phenomena is lost. Even if a geographic component is brought in by showing the developments in the various regions separately, it is still far more difficult to draw spatial conclusions from this

complex image than if the same information had been mapped. Of course, there are various possibilities for the design of a diagram, e.g. a line diagram or histogram, a representation with index values or a logarithmic presentation. Some possibilities are shown in Figure 3.10. Should a map be selected, then one first has to select the aggregation level on which the information has to be depicted: on the level of statistic enumeration areas, municipalities or even higher-order areas. The resolution (level of spatial discrimination) at which the information is presented depends on the space that is available for the map and the goals of the map author. On the level of local enumeration areas or municipalities, one would be able to show the local trends, concentrations and dispersions. On the basis of aggregation of municipalities or at a county-level, only regional patterns can be shown.

After the selection of the aggregation level of the data, a choice can be made between an absolute or relative representation. The results of both choices may be contradictory: a map of the absolute unemployment in the Netherlands shows a concentration in the central Randstad area, whereas the unemployment problem seems to lie in the marginal areas in the south, east and north of the Netherlands when judging from the image of the relative unemployment (Figure 3.11). Selection of either absolute or relative representation also implies the selection of a specific cartographic method, that is for proportional symbol maps or choropleth maps, respectively. The same information could also be presented as an isoline map, a grid map or a map with diagrams. This will also depend on the ulterior motive or objective of the communication process.

Supposing data are available relating to the participation of a city's inhabitants in public transport, and that it is known through questionnaires or otherwise how many passengers were transported over each stretch of the bus routes between bus stops, where the passengers boarded and descended again, and where they came from and went to. Similar data have been collected for a number of major cities. If one wanted to show the existing spatial relationships between the inhabitants, then the movements by public transport from each part of the city could be shown through sets of arrows. If the transport authorities were more interested in the routes followed by the inhabitants to get to their destinations, these could be plotted, and the frequency with which they were used could be shown by proportional widths (not indicated in Figure 3.12). By comparing origins

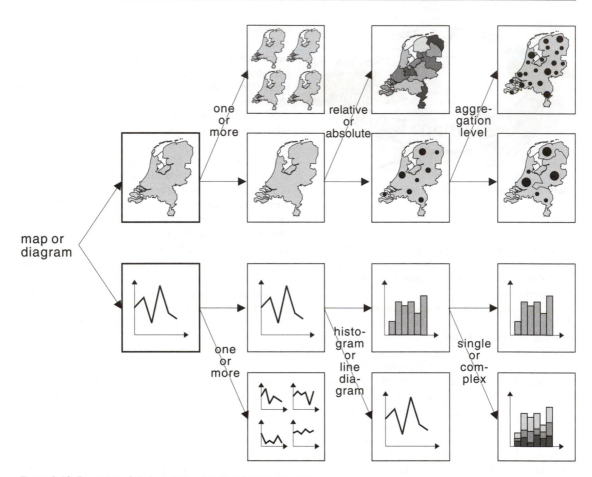

Figure 3.10 Examples of choices during the visualization process

and destinations of all the trips recorded in the questionnaires, the places where people changed buses could be traced, and a hierarchy could be portrayed, showing these bus stops with circles proportional to the number of changes.

Accessibility of the public transport network could be shown by portraying the number of bus stops per regular area unit. The data could be aggregated for the various wards the city would be divided in, showing the volumes of passengers travelling between the major wards by proportional arrows. By comparing the destinations of the passengers from each city ward with the public transport timetable, the average travelling time could be determined, and portrayed through isochrones. The degree of participation of city inhabitants in the public transport, computed either as a percentage of all inhabitants, or as the number of people travelling by bus per areal unit,

could be expressed three-dimensionally. So depending on the actual aims or objectives, the database built up from the questionnaires could be used for the production of quite different types of maps, each conveying the answers to different questions.

The possibilities for its reproduction and the available financial resources determine whether the map will be made in black and white or in colour, whether it will have to be very much generalized because it is printed on poor-quality (newsprint) paper and whether for instance map lettering may be applied in various styles and sizes. The manner of distribution, the way in which it will be documented and the management in libraries will determine who will see the map, and therefore who will be able to use it. The role of publishers is essential here, as it is usually they who are able to control the distribution, through assigning specific prices to the cartographic products,

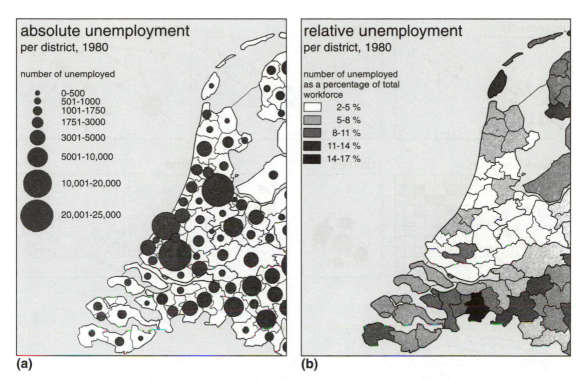

Figure 3.11 (a) Absolute and (b) relative presentations of the same information (unemployment data)

through adding ISBNs to the maps, through the use of their distribution outlets and channels, through marketing campaigns and finally through the distribution of review copies to cartographic journals. The map's potential for use will also be determined by the skills of the users and their foreknowledge about the theme that is mapped out. Of course, the first requirement here is that map user and cartographer speak the same graphical language, that they use a common set of symbols or associations, so that the signs in the map are understood by the map users, and so that they can communicate. The addition of legends is another essential aspect of map documentation and it is not always automatically generated, not even in the newest GISs.

Map use is also the subfield where the cartographer does research on the map's effectiveness. There are currently two different attitudes to improving cartographic communication: one is to improve map reading expertise by user training and the other entails improving the maps. Both routes, of course, have to be followed simultaneously. In many countries (cf Great Britain with its map use standards) high

school students during geography lessons are confronted with standard tasks they should be able to perform with the help of maps in specific grades (e.g. between the ages of 13 and 16 pupils should normally be able to correlate information on two or more thematic maps in an atlas such as relief and vegetation) (Boardman, 1983).

The other route, that leading towards improved maps, has been boosted recently by the advent of the computer. It has not only allowed cartographers to experiment freely with map design options, it has also allowed them to customize maps, i.e. to adapt them to specific requirements. A good example is the hydrographic chart. These are rather complex documents as they have to cater for many different types of map use. They contain information about shipping hazards, depths (isobaths, soundings, wrecks, reefs), currents, tides, navigational lights and buoys. They contain several grids for locating purposes (geographical coordinates, decca grids), boundaries, one-way shipping lanes in densely navigated waters, warnings, munition dumping areas and drilling platforms, and also contain information on the nature of

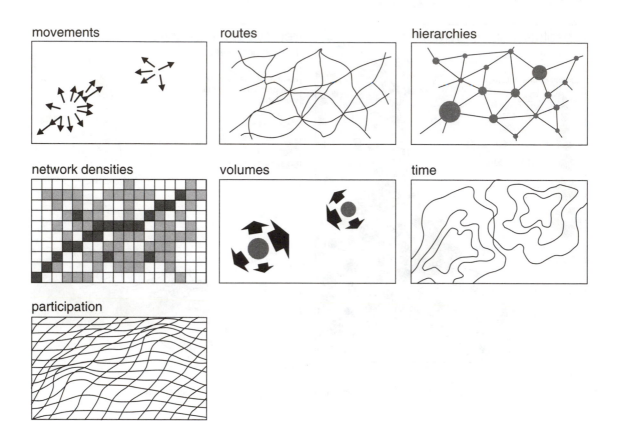

Figure 3.12 Visualized results of changes in definition of attribute data from a public transport database

the underwater soils, etc. When navigating one never needs all this information at the same time, and some of the information categories one never needs.

Now that electronic charts are emerging, with all these layers of information categories, one has the choice to visualize only those layers one needs for a specific purpose, thereby avoiding the clutter of symbols on paper charts. If one's ship draws only 3 m, then all the depth information beyond the 3 m isobath can be omitted. During daylight one would not need information on navigational lights; during night-time one would only need information on the location of those buoys that also have lights. So one is able to adapt the map displayed on the monitor screen to the specific navigational requirement of a specific time and place.

In order to check a map's effectiveness in conveying spatial information, test subjects are asked to perform tasks on the basis of the map. When testing out various versions of a town plan, test subjects could,

for instance, be asked to locate addresses, to describe the shortest route from one address to another, or to describe the optimal link through public transport between two locations. The use of maps with the same informational content but with different contrasting colours, type sizes, degrees of generalization or grid reference systems might result in different performances regarding the time it took to answer these test questions correctly.

On the basis of map use tests like these, the best possible design for answering specific tasks can be selected. However, one not only has to take into account the functions the map has to perform, but also the circumstances in which the spatial information will be used (constraints in space, lighting, etc.) and the constraints imposed by the target audience (age, school experience, etc.) which might result in a specific colour selection, and affect the degree of map complexity and the wording of the marginal information.

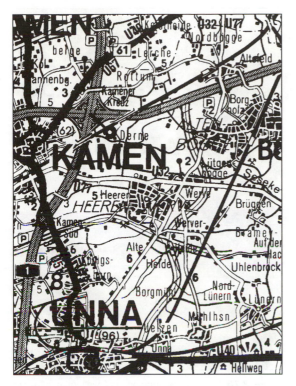

Figure 3.13 Topographic map used for orientation purposes

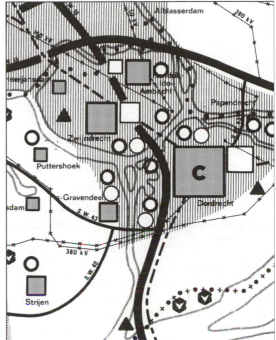

Figure 3.14 Physical planning map (Courtesy *National Atlas of the Netherlands*, 1st edition)

3.4 Map functions and map types

The most important function of maps is probably the function of orientation or navigation. In any case, most of the maps the general public comes across, with the exception of weather charts, are produced as an aid to orientation and navigation (Figure 3.13). People use orientation maps (e.g. road maps, topographic maps, charts) for getting from one place to another along a selected/plotted route, and want to be able to check with help of the map/chart whether they are still 'on course' during their trip.

Maps that are used for physical planning take second place to orientation maps, although it would be the other way around if it were the number of different maps available rather than the total number of printed copies that were the issue. Physical planning maps are maps that inventorize the present situation, maps that define development processes, and maps that contain propositions for a future situation, e.g. future land use. Generally, alternatives have been

made for such plans as well, which are offered to the public in the form of planning permission applications. Before a development plan or regional plan is codified in its ultimate form, hundreds of these physical planning maps will have been made (Figure 3.14). Also directed towards the future are maps that show forecasts: certain developments from the past are, on the basis of the development pace that is to be expected, extrapolated into the future. This applies to weather charts, as in Figure 1.17, but may also hold good for other forecast maps, such as those that show the expected spread of insects or diseases.

Maps used for management tasks or monitoring purposes are generally large-scale maps that are manufactured bearing in mind the management and maintenance of objects, e.g. roads, railways, forests, dikes, canals and airports. In the Netherlands detailed maps of the sea dikes used to be produced, with 1 m isohypses. After every major storm the dikes would be photographed again from the air, and these images would be photogrammetrically pro-

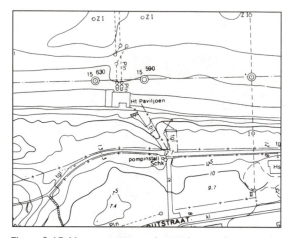

Figure 3.15 Management map (coastal protection map, the Netherlands)

Figure 3.16 Detail of an educational wall map of Africa (Courtesy Westermann)

cessed and turned into new maps, to be compared with the old ones, in order to ascertain whether the coastlines would have been damaged by sand or dunes being swept away. Where this was the case, the sand would be replenished as soon as possible. These maps used to be printed first, but the speed of processing the new data is nowadays often so fast that reproduction is often done by diazo printing (Figure 3.15), by plotting from a GIS, or by new printing on demand techniques.

For educational objectives, special purpose map material has been produced since around 1750; school atlases, wall maps (Figure 3.16) and workbooks provide the pupils with a spatial frame of reference in order to be able to understand national and world-wide developments.

Another map function is codification, e.g. showing the legal situation as regards property rights. In continental European countries, and indeed in an increasing number of countries, cadastral maps are being produced that have this function of codifying land ownership.

Different categories of maps are based on these main map functions (orientation, physical planning, forecasts, management and education), and on others which will not be discussed here. They should be discriminated from map types, which are groups of maps designed according to the similarity in the specific methods used, such as the chloropleth method or the isoline method.

Maps may also be subdivided according to themes: physical planning maps, town plans, weather charts, geological maps, population maps, language maps, etc. From a map design or a map use point of view

it is no use discussing these categories of map separately, since for different map themes, identical design, representation or interpretation problems may occur. So in Chapter 7 the design problem will be the guideline of the discussion, rather than the theme that is depicted. The division into map categories as such is not of importance either for the discussion of the possibilities for the derivation of information. Map categories are discussed along with map use in Chapter 10.

Further reading

Morrison, J.L. (1980) Computer technology and cartographic change. In D.R.F. Taylor (ed.) *The Computer in Contemporary Cartography*. New York: Wiley & Son.

Rhind, D. (1980) The nature of computer assisted cartography. In D.R.F. Taylor (ed.) *The Computer in Contemporary Cartography*. New York: Wiley & Son.

Robinson, A.H. and B. Bartz Petchenik (1977) *The Nature of Maps*, Chicago: University of Chicago Press.

CHAPTER 4   GIS applications: which map to use?

4.1   Maps and the nature of GIS applications

To solve problems in the fields of earth and social sciences, maps are often indispensable. Frequently, the researcher will have questions related to the nature and coverage of some particular phenomena. Finding answers will involve questions concerning what map type to use and which map scale to use. Map types and scales have been covered in Chapter 3, and one might be inclined to think that in a digital environment map scale is no longer relevant, as it is possible to zoom in or out on the data at will, once they have been digitized. This misconception has to be countered here, as the scale at which spatial data have been digitized does indeed matter. Keywords here are clutter, resolution and being representative. When map data have been digitized at a specific scale and then rendered on the monitor screen at a much smaller scale, spatial data on the map may be too dense, so that the overview is lost. Particular objects with a detailed character, such as coastlines, might appear as a cluttered line when reduced to a smaller scale. On the other hand, line elements that have been digitized at a specific scale and are rendered much larger on the screen, may no longer be representative or characteristic, as straight lines and sharp angles might be acceptable at a small scale but will look unnatural at larger scales. Moreover, an increase in scale might lead to a map image with too little topographic information. It is therefore wise either to use boundary files that have been digitized in a similar scale range as the one needed, or to generalize existing files (for generalization, see Chapter 5).

4.2   Cadastre and utilities: use of large-scale maps

4.2.1   Cadastral maps in use

Land has been registered and mapped since ancient times because it is vital to humankind. We all somehow depend on land for our living. The cadastre, an information system that uses the land parcel as the basic geographic unit to register land ownership, was revived in continental Europe by Napoleon, who used it to collect taxes. The cadastral map plays an important role in this system, since it shows boundaries that define the ownership (Figure 4.1). Cadastral maps are large-scale maps normally at a scale between 1:1000 and 1:2000. They sometimes include the outlines of important buildings next to legal boundaries for orientation purposes. The parcels themselves are often numbered, and these numbers form the link to other components of the cadastral information system. These could be the registries of landowners or the original survey data.

The nature of the relationship between the cadastral map and the topography differs, depending on the country's cadastral system. In some countries, such as Great Britain, the general boundary system is in use. Here the cadastral boundaries coincide with topographic objects; in other words, the boundaries are visible in the terrain. In the United States and the Netherlands it is not necessary that cadastral boundaries coincide with topographic objects. However, in the United States they are marked in the land by pegs, while in the Netherlands the original survey

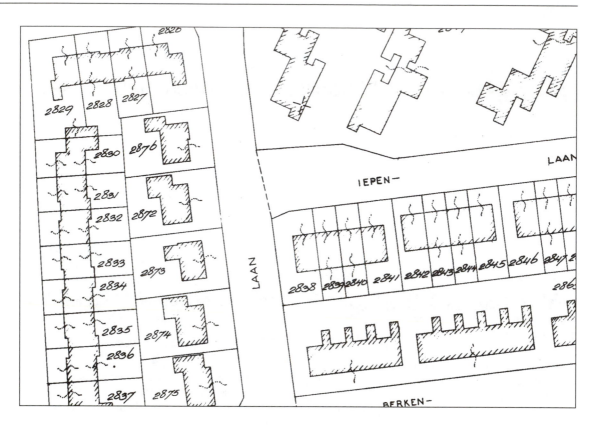

Figure 4.1 Cadastral map of the Netherlands (1:1000): a detail showing ownership boundaries, houses and parcel numbers. The 'snakes' indicate that houses and surrounding areas belong to the same parcel

data have to be reconstructed to localized cadastral boundaries in the terrain.

Cadastral maps have had many uses during the last few centuries. Among these are land reclamation (in the 16th century in the Netherlands), evaluation and management of state land resources (in 19th century United States and Canada), land redistribution and enclosure (in 17th century England), colonial settlement, and land taxation. Nowadays, similar tasks are executed by what is called a multipurpose cadastre, a parcel-based land information system (Figure 1.7). Dale (1991) provides many examples of the potential content of a multipurpose cadastre: data that define the parcels, land tenure data, data on land value and land use. Data on buildings, infrastructure, population and administration can also be included.

The process of rural land development is an example of the manner in which the cadastral map is used in a GIS environment (Figure 4.2). In some rural areas the need to change the current historic pattern of land parcels arises because the parcel pattern no longer fits modern agricultural requirements nor an economic approach to the use of land. In a land reconsolidation process, land ownership is a vital factor, but not the only factor. The land in the study area will not be of the same quality in all places and, because of this, reallocation based on ownership only would not be fair. The reallocation is therefore executed on the basis of the value of the land. Land value is determined by combining the ownership data, the topography and a soil map. In a GIS, the value map is created by an overlay operation. The land value map is combined with a physical planning map that shows the new layout of the land (drainage system, road pattern, etc.). A model linked to the GIS will then, in an interactive iterative process, define the new parcels and ownership. The result will be a new cadastral map and often a new topographic map as well.

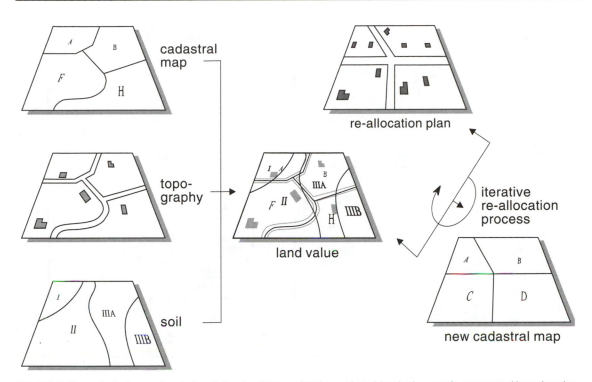

Figure 4.2 The cadastral map at work: its role in a land reconsolidation project. A land value map is constructed based on the cadastral map, a topographic map and a soil map. In an iterative process the land value map is combined with a re-allocation plan to create the new cadastral map

4.2.2 Utility maps at work

The function of maps in the environment of utility companies is to keep an accurate and up-to-date record of their infrastructure for maintenance and planning purposes. Examples of utility networks are those of gas, water, sewerage, electricity, telephone, cable television, and water. Next to geometric information related to the location and depth of the pipes and cables, the maps register attribute information such as a pipe diameter, construction material, age and capacity. The types of maps needed are large-scale maps, that would for instance be used by a field engineer to locate a plant for maintenance, and small-scale maps to plan future demand.

The large-scale maps in use are often those supplied by official mapping organizations. They are used as a base on which to map the pipes and cables. When the utility companies have to survey their own maps, they show only selected topography. For instance, only the distance from the front of a building is needed to locate the cables or pipes under the pavement. Figure 4.3 shows an example for water. The coding along the pipes refers to their attributes. The location on the map is not exact. However, to avoid digging in the wrong place, with the possibility of damaging other cables or pipes, or even finding no cables or pipes at all, survey data are included as well, as can be seen in the detail in Figure 4.3. Maps of dense urban areas can become very cluttered because of the high density of lines and text representing cables or pipes. In these situations the cables and pipes are often found above each other in a special ditch. In a GIS environment a solution to these graphic problems is the application of a layering technique. Although, in practice, the utility companies can solve their automated mapping and facility management problems, it is very likely that a water board will still need to consult an electricity company before starting to work in a particular area. Information is exchanged, often via local authorities, to keep each other informed on the whereabouts of cables and pipes. This exchange is not only related to the location of the network but also to the planning of works

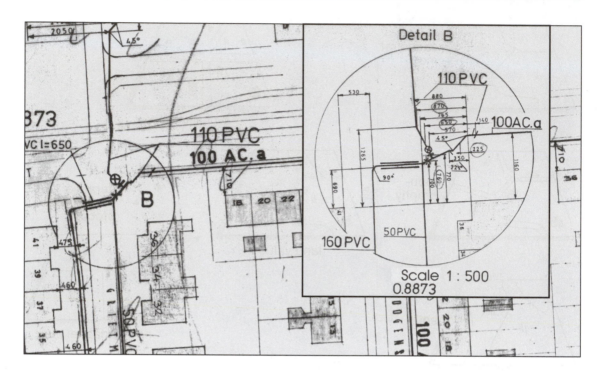

Figure 4.3 A utility map at a scale of 1:1000: water conduits and pipes

to avoid a street to be successively opened and closed by different utility companies in a single month.

Schematically designed small-scale maps are used in the planning process of utility companies. Such maps contain only the network lines and selected attributes, and are used to plan future extensions or to play a role in emergency simulations. In the planning process, maps with geological and soil data are also used to determine the character of the subsurface for the underground pipes and cables. A database with customer information is also vital to the utility companies because it provides them with knowledge on the nature of the customers' connections as well as the addresses to send bills to.

Figure 4.4 demonstrates the role of utility maps in a telecommunication company's GIS. The monitoring system, tracking the status of the network, signalled a problem in a certain circuit, which damaged a certain part of the network. On an overall scale the system tries to reduce the damage by rerouting most of the traffic along other paths of the network. Only local traffic is affected now. Meanwhile, field engineers have localized the trouble and quickly produce a large-scale map that includes the necessary geometric and attribute information. A call to the local

authorities provides them with a map that shows all other pipes and cables near the trouble spot. From the database a material list is generated of those components that may be needed to repair the damage.

4.3   Spatial analysis in geography: use of small-scale maps

4.3.1   Socio-economic maps

Themes represented in socio-economic maps are often derived from statistics related to topics such as census data, infrastructure, housing, employment, trade and agriculture. Figure 4.5 shows two examples. The map on the left shows how absolute values (here fish catches in Africa) are represented by differently sized circles. Interesting to see in this map are the high values for land-locked countries such as Chad and Uganda. Studying the map in more detail reveals that those countries have large inland lakes (Lake Chad and Lake Victoria). It is a good example of the need to add useful topographic information to

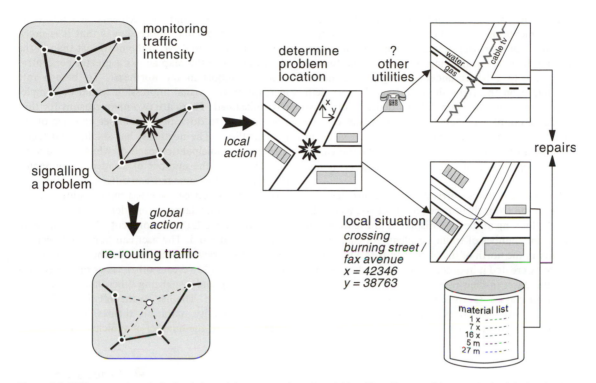

Figure 4.4 Utility maps at work; their role in maintenance and repair activities. Signalling a problem results in global action (re-routeing traffic) and local action (preparing and executing repairs)

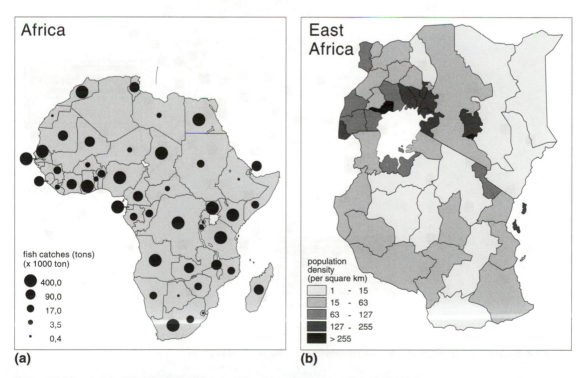

Figure 4.5 Mapping statistics: (a) fish catches in Africa; (b) population density in East Africa

thematic maps. It is not only necessary for orientation, as mentioned in the first section of this chapter, but also for a better understanding of the map. Adding the rivers to the map of Africa would further contribute to a better understanding of the theme. The figure on the right shows a typical choropleth map depicting the population density of East Africa. Examples of socio-economic maps can be found throughout this book. Chapter 7 gives in more detail the characteristics of the different cartographic options available to map socio-economic data.

To interpret maps correctly and retrieve relevant information from them is not always easy, as can be seen by looking at the map displayed in Figure 4.6. It shows the number of traffic accidents for each municipality in the north of the Netherlands. An insurance company created it in order to help it decide on regional differences in their rates. The data have been split into four classes. Graphically this results in squares that can have four sizes. The smallest square, for instance, is drawn in those municipalities that

counted between 1 and 100 accidents. When looking at the map one's first impression is that it is much more dangerous to drive in the northeast than in the northwest since the map shows a greater concentration of symbols in the northeast. To base a rate increase on this visual impression would be completely unjustified. There are two main reasons for this. The first has to do with the number and size of the municipalities. The northeast has smaller and therefore more municipalities, each of which gets a symbol. However, the number of accidents represented by squares of the same size can differ considerably, as a comparison between selected municipalities shows when the original data are considered. When working in an interactive GIS environment, the original data are usually at hand. The user can point and click on the map to retrieve the data behind the map. It is also possible to quickly create other views on the subject by combining and visualizing other data or changing the classification. The list on the left shows that, for each municipality, data on themes like the total length of roads, the number of vehicles per kilometre

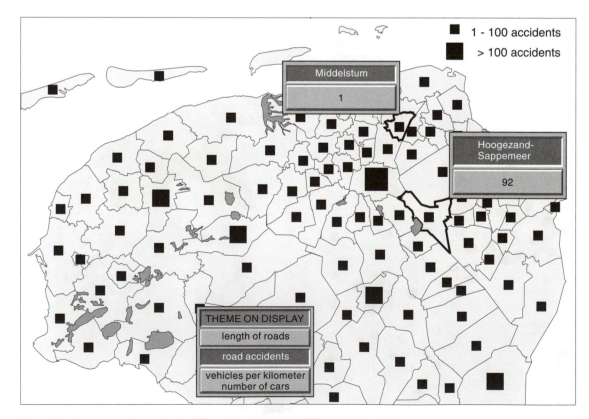

Figure 4.6 Socio-economic maps and GIS: road accidents in the Netherlands in 1990

road and the road accident's toll per 1000 inhabitants are available. An accountable decision on local differentiation in insurance rates can only be made when all relevant and interrelated information is studied properly. To do so, insurance companies will have formulas or models where each parameter is given a weighting.

4.3.2   Environmental maps

Environmental maps are created to gain a better understanding of the earth's natural resources. Some of these maps are inventories related to vegetation, soil, hydrology, geology, geophysics and forestry. Others are related to the use and misuse of these resources, such as maps showing water, air or soil pollution. Inventory mapping programmes are often extensive and time consuming, especially in remote areas. To get up-to-date information on land use for those areas through traditional fieldwork techniques is difficult. Here the application of remote

sensing techniques can be very helpful. Figure 4.7 provides an example. In a study for a rural development plan in the Koundara area in Guinea, a SPOT image at a scale of 1:100 000 (Figure 4.7a) was used to execute an inventory of farmland use and soil potential. It should identify the current agricultural land use as well as constraints on and potential of the remaining land. The resultant map (Figure 4.7b) indicates the fallow land, the land currently cultivated, the uncultivated but potential farmland, the uncultivated but unsuitable land, and the forest areas. It is an example of qualitative mapping.

For the urban area of Athens (Greece) a study has been carried out to judge the magnitude of air pollution (Koussoulakou, 1990). The study resulted in an air diffusion model which could be used to predict the effects of different air pollution parameters. Maps and GIS played an important role in the model, which was capable of modelling a diversity of spatial data. The model was calibrated with real-world data in order to be able to use it later to predict the level of air pollution under different circumstances.

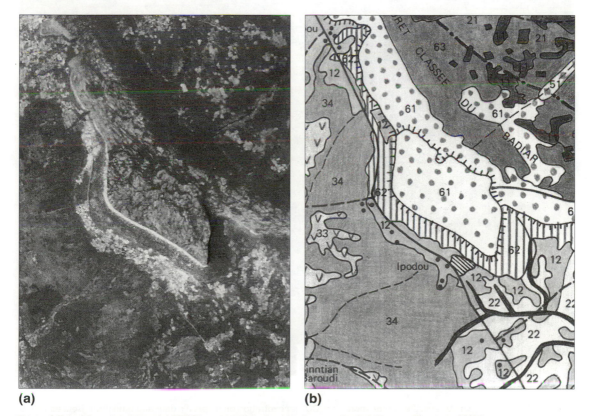

(a)   **(b)**

Figure 4.7 Land use inventory: (a) SPOT images; (b) derived land use and suitability map (Courtesy CNES)

Parameters in the model were related to the topography (Athens is located in a bowl bounded by mountains and the sea), meteorology (wind, temperature, precipitation), urban conditions (land use, population, traffic density), and sources of emissions (industries, traffic, central heating). The individual parameters were each mapped in separate analytical maps using different thematic mapping techniques. The model itself, combining all parameters, resulted in several animated maps that showed air pollution concentration developments above the city during the day.

Figure 4.8 is an example of an environmental map, displaying the concentration of carbon monoxide (CO) in the area. The quantity of CO is shown by tints of grey. Each represents a certain range of the data. The map itself is a single frame from a cartographic animation that shows the change in CO air pollution during 13 April 1985. The clock on the right shows the current time. Below the map an interface allows the user to interact with the animation, and move forwards and backwards through time. More details regarding cartographic animations can be found in Chapter 9.

4.3.3 Spatial, thematic and temporal comparisons

While working with spatial data in a GIS environment it is very like that one will have to deal with three basic query types: 'Where?', 'What?', and 'When?'. In a spatial analysis operation the queries will result in the manipulation of the spatial data's geometric, attribute or temporal components, separately or in combination. However, just looking at a

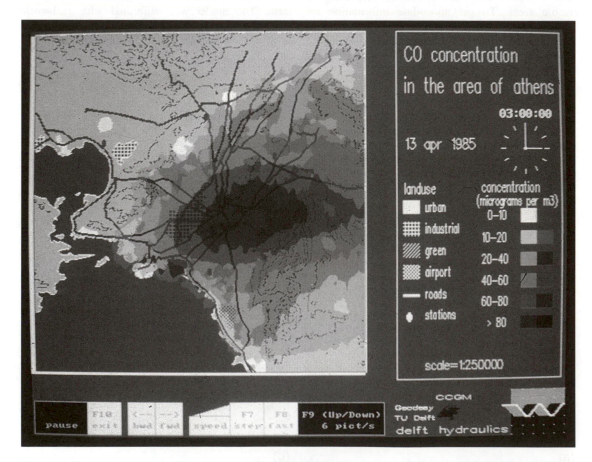

Figure 4.8 Air pollution in Athens, the carbon monoxide (CO) concentration on a specific date and time (from Koussoulakou, 1990)

map that displays the data already allows one to evaluate how certain phenomena vary in quantity or quality over the mapped area. Often one is not just interested in a single phenomenon but in multiple phenomena. For some aspects analysis operations are required, but sometimes even a visual comparison will reveal interesting patterns for further study. Spatial, thematic and temporal comparisons can be distinguished. A spatial comparison means looking at different areas at the same scale, to see if certain patterns correspond or differ. A thematic comparison means looking at maps displaying different themes of the same area to see if the spatial distribution of the themes is similar or different. Temporal comparisons are executed by studying views of the same area for different dates. The next subsections will look at each of these comparisons in more detail.

Comparing spatial data's geometry component

Comparing two different areas seems to be relatively easy while focusing on a single theme, e.g. hydrology, relief, settlements or road networks. However, to make a sensible comparison the maps under study should have been compiled according to the same methods. They should have the same scale and the same level of generalization or adhere to the same classification methods. If two areas are compared in order to get an impression of the population density based on the settlement density, both maps should show the same type of settlement. They should both show those settlements with, for example, more than 10 000 inhabitants. If one is comparing the hydrological patterns in two river basins the individual rivers should be on the same level of detail with respect to generalization and level of branches.

In Figure 4.9 the deltas of the Rivers Rhine and Scheldt (the Netherlands) and the Rivers Neuse and Roanoke (North Carolina/Virginia) are shown. At first glance both areas look quite different. When one changes the orientation of the Neuse/Roanoke delta, this view suddenly changes. The shapes of the coastlines clearly have something in common. To ease the comparison the names in both maps have been left out. The links between the towns in the Neuse/Roanoke delta seem to be there just to connect the harbour with the hinterland. Links in the Rhine/Scheldt delta are more regular. On the basis of the more numerous settlement symbols, this area seems to be much more densely populated. The hydrographic pattern in the Roanoke/Neuse delta area is apparently much more of a barrier than in

the Rhine/Scheldt delta area. So simply by comparing a map of an unknown area with one of a known area (produced to the same specifications) it is possible to learn quite a lot about otherwise unfamiliar areas. However, this is only possible when one has homogeneous geographic maps available. In the small-scale map range the 1:1 million International Map of the World (now also in digital form) and the 1:2.5 million Karta Mira are available (see also Chapter 5). Standardization of their legend and the uniform level of generalization make these maps very suitable for small-scale comparisons.

The comparison in Figure 4.10 is of a different nature. To study the similarity between the estuary of the River Thames near London and the Solent near Southampton and Portsmouth, not only the orientation of the area has been changed; both areas have been transformed into simple geometric shapes as well. This example has been adapted from Cole and King (1968). To study the industrial development patterns of both areas (the docklands) the orientation has been altered so that comparisons are easier. The upstream regions of both rivers are now at the north and the open sea at the south of both maps. For the Thames this means a 90° rotation. In addition, the Solent was also mirrored and its scale enlarged. To minimize the influence of the shape of the river banks and islands the map was further generalized, as can be seen in Figure 4.10(c). Actually this type of generalization is called schematization, because of its rigorous approach. Comparing both schematic maps, it can be seen that the main urban areas can be found upstream. The docklands are to be found there as well. Both areas show oil refineries near to the open sea (large oil tankers), and a resort area very close to the open sea.

Comparing the attribute components of spatial data

If two or more themes related to a particular area are mapped according to the same method, it is possible to compare the maps and judge similarities or differences. However, not all mapping methods are easy to compare. Choropleth maps are the most simple to compare, at least as long as the administrative units are the same in both maps. To be able to compare qualitative maps (e.g. Figure 4.7b) they must be converted into for instance, isoline maps. Comparing isoline maps is a well-established technique. A classical example is the study by Robinson and Bryson (1957),

(a)

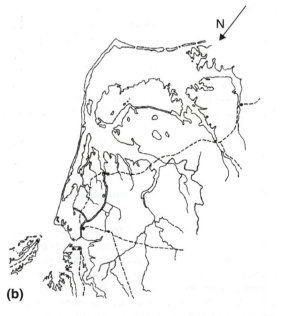

(b)

Figure 4.9 Geographic comparison: similarities between the Rhine Delta (right) and the Cape Hatteras area (left): (a) both areas north oriented; (b) Cape Hatteras area rotated

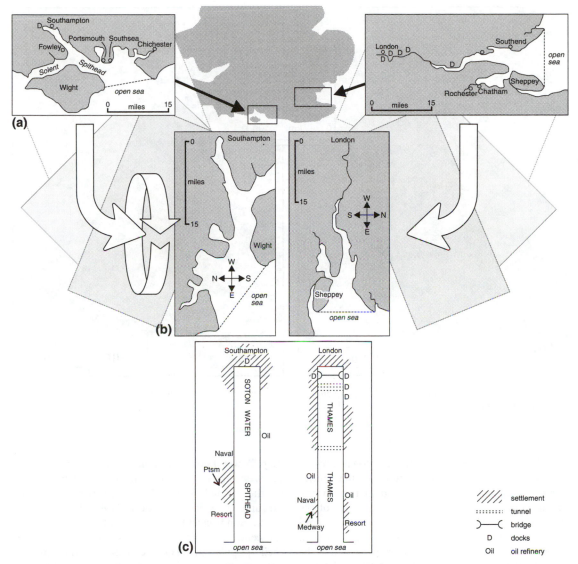

Figure 4.10 Topological comparison looking at the River Solent and the River Thames (after Cole and King, 1968): (a) both areas in correct orientation and scale; (b) after manipulation of orientation and scale; (c) both areas after schematization

who compared the precipitation and population density maps of America's Midwest. The importance of their study is partly due to the fact that the authors considered the accuracy of the information during the comparison. Comparing isoline maps is executed by measuring and comparing values in each map at the same location. Figure 4.13(c) shows an example of this approach. Comparing maps with point symbols can only give a rough idea of similarities or differences.

The maps of southeast Britain in Figure 4.11, based on census data, show the distributions of children and persons over retirement age. Both age groups are mapped relative to their average value for the whole country. On average, children under 15 years of age make up 24% of the population and those retired 16%. The distribution of the children is closely linked to the distribution of younger married couples. From the maps it can be seen that they are concentrated in the suburban areas around

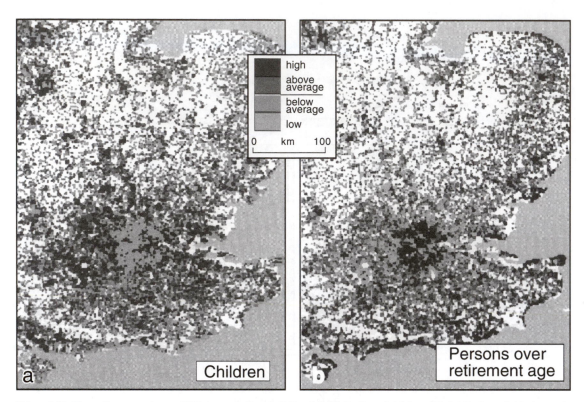

Figure 4.11 Thematic comparison of British population statistics: (a) distribution of children; (b) distribution of elderly people (Courtesy HMSO, *People in Britain, A Census Atlas* 1980)

most large cities as well as in the new towns around London. They are underrepresented in the central London area, along the coast, and in rural areas. The map in Figure 4.11(b) is a mirror image of the children's map. The elderly people are obviously concentrated along the coastline, especially in the south. People over retirement age are overrepresented in the central urban reas and most rural areas as well. Some of these concentrations are popular retirement areas, like the south coast, while other concentrations, such as those in the rural areas, are due to the younger people leaving these areas.

Comparing the temporal components of spatial data

Users of GIS are no longer satisfied with analyses of snapshot data but would like to understand and analyse whole processes. A common goal of this type of analysis is to identify typical patterns in space over time. Change can be visually represented in a single map. A well-known example are the maps displaying the westward movement of the centre of population of the United States. It moved from Maryland (around 1800) into Missouri (1990). The centre of population is calculated by summing for each census the centre coordinates of all counties, weighted by their population. The trend shows the growing importance of the west coast population. Understanding the temporal phenomena from a single map will depend on the cartographic skills of both the map maker and map user, since these maps tend to be relatively complex. An alternative is the use of a series of single maps, each representing a moment in time. Comparing these maps will give the user an idea of change (see, for instance, the maps in Figures 1.11, 4.12 and 4.13). The number of maps is limited since it is difficult to follow long series of images. Another, relatively new alternative is the use of dynamic displays or animations. A well-known example is the animation shown during the daily weather report on television presenting the change in the atmo-

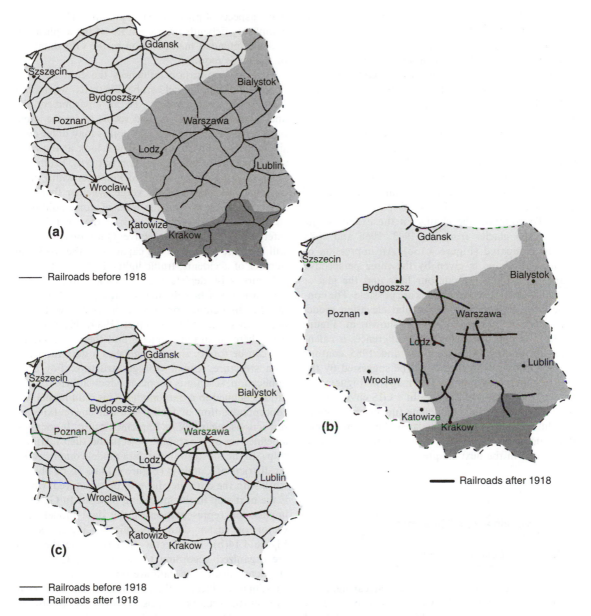

Figure 4.12 Temporal comparison of the development of the Polish railroads (after Grote Bosatlas, Courtesy Wolters-Noordhoff): (a) railroads built before 1918; (b) railroads built after 1918; (c) full railroad network

sphere's meteorological conditions. The Athens map in Figure 4.8 is another example. Chapter 9.4 will explain more about cartographic animations.

Figure 4.12 shows how line patterns can be compared. To be able to do so, change has to be emphasized. The upper map in the figure is a representation of the Polish railroads at the beginning of this cen-

tury. The pattern reveals the political boundaries at that time. The lower map shows today's railroad network, which is clearly more dense than the one in the upper map. However, to understand the changes, a map showing the changes has to be produced. From this map, which also includes the former boundaries between Austria–Hungary, Germany and Russia, it can be understood that the railroads added were built to erase the boundary pattern and emphasize the importance of the new economic centre, Warszawa.

Interest in the behaviour of glaciers under today's environmental conditions led to the Greenland Ice Margin Experiment (GIMEX; Roelfsema *et al.*, 1995). Part of this research project consists of monitoring change. One of the glaciers that researchers focused upon was the Leverett Glacier on Greenland's east coast. By building three digital elevation models based on aerial photographs from 1943, 1968 and 1985, glaciologists expected to gain an impression of change in the volume and shape of the glaciers. Some results are displayed in Figure 4.13. To compare the change in the extent of the glacier, the glacier fronts in 1943, 1968 and 1985 can be compared (Figure 4.13b). An impression of changes in shape is given by the three perspective maps. By adding the contour lines and the glacier margins, absolute comparisons are possible. The contour lines derived from the digital elevation model can also be compared, as is shown in Figure 4.13(c). A map displaying relief differences is calculated by adding both the 1968 and 1985 contour maps. Growth or decline could be suggested by the use of different symbols. The maps in Figure 4.13 give an impression of change. In a GIS environment with extensive surface modelling functionality, more absolute values can be calculated. Other map products derived from digital elevation models will be elaborated in the next chapter.

4.4 Working with digital data

4.4.1 Modelling the world

Making today's maps without digital cartographic models is almost unthinkable. As explained in Chapter 1, digital cartographic models are derived from digital landscape models (Figure 1.3). It is the content and structure of these models which determines the possibilities for querying the data, and for defining a digital cartographic model needed to draw maps as required. The content of the digital landscape model itself is defined by selections from reality. Obviously reality as experienced outside cannot be incorporated in a model as a whole. Selectivity is necessary to keep the model workable. In the framework of an application one will try to process as many aspects of phenomena as possible. The model's complexity depends on the nature of the application and the intended manipulation in the GIS database or the maps required.

How an application influences the contents of a digital landscape model can be illustrated when considering the road concept. To an environmentalist and a traffic manager a road seems to be the same object. However, the two viewpoints may differ considerably. An environmentalist will look at the road as a barrier to wildlife migration patterns. From this perspective he or she wants to know its width, how busy it is, the width of the verge, whether there are any crash barriers, and the level of noise pollution, etc. A traffic manager will look at the road from a safety and transport perspective. Questions he or she will have are related to the capacity of the road, the number of accidents, traffic lights, flyovers, etc.

Figure 4.14 depicts the modelling process. The illustration is a broadening of Figure 1.3. The steps in the modelling process correspond with the approach suggested by Peuquet (1984). Next to reality, where geographic objects and their characteristics can be found, she distinguishes a data model, a data structure, and a storage structure. For each application, selections are made which adhere to conditions defined in a data model. This data model is a conceptualization of reality without conventions or restrictions regarding its implementation. It contains a defined set of geographic objects and their relationships. In the illustration the data model comprises the six districts of Cumbria (Figure 4.14a). To be able to work with the data they have to be structured, which is the next step in the modelling process. In GIS this implies the representation of the data model in a vector or raster data structure as illustrated in Figure 4.14(b). In a vector data structure the data are organized according to the objects. Geometric characteristics of the data are represented by sets of coordinates which, in the map image, are connected by lines (the vectors). Labels link the attributes to the geometry. In a raster data structure data are organized on the basis of a spatial address. The geometry is represented by the location of grid cells. The address of the cells links the attributes.

The choice of a suitable data structure should be determined by the application. However, in practice one is restricted to the data structure implemented in commercial software packages. Still, when judging the data structure one should at least consider points such as completeness, efficiency, lineage, versatility and functionality. Completeness

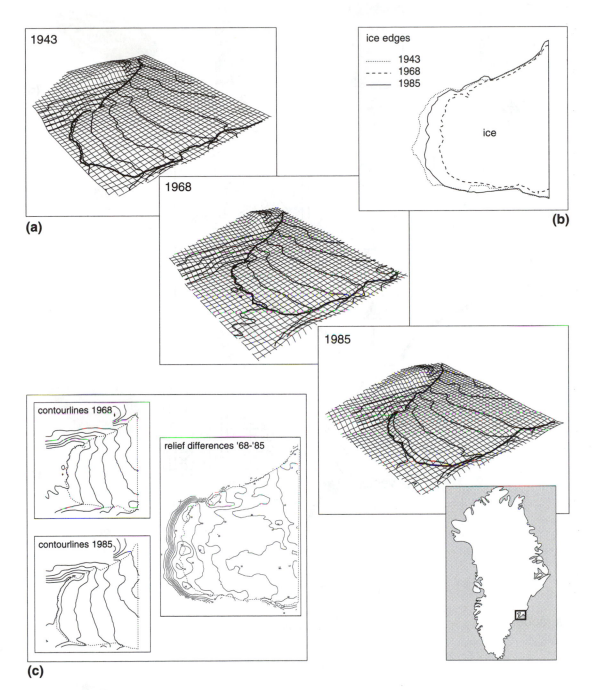

Figure 4.13 Decline and growth of the Levett glacier in Greenland (Courtesy Roelfsema *et al.*, 1995): (a) the changing glacier in a perspective view; (b) the ice-edges through time; (c) differences between the 1968 and 1985 ice-cover

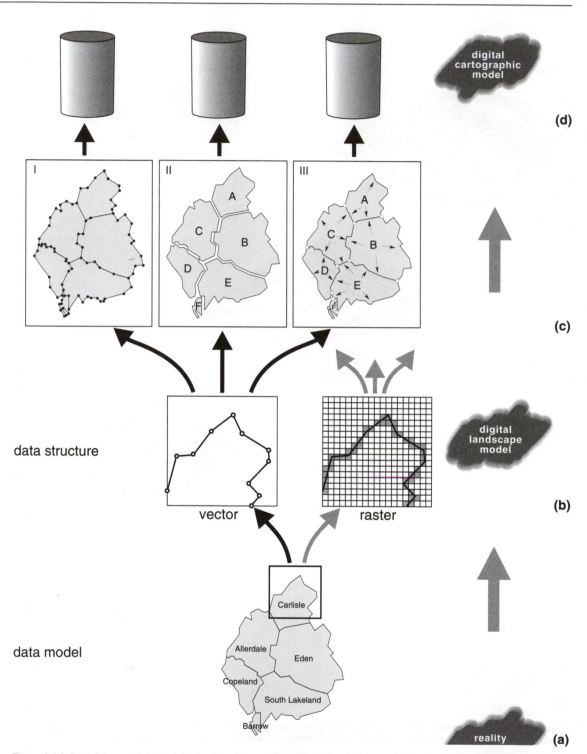

Figure 4.14 Organizing spatial data (a) selections from reality are based on a data model; (b) the selected data are often structured in a vector or raster format (digital landscape model); (c) the nature of the data structure defines the query level, and (d) determine what maps can be drawn (digital cartographic model)

refers to the possibility of representing all selected data, efficiency refers to data accessibility, lineage refers to the way in which the data were collected, and versatility refers to the possibility of adapting the structure to new circumstances. Functionality refers to the operations that can be executed with the data; in other words, to the kind of query that can be answered by the structure. The query level will define the structure's suitability for certain tasks. This is illustrated in Figure 4.14(c) by the three examples of some possible vector data structures in relation to its cartographic use. Map I represents just the individual lines, map II represents the areas as well, and in map III, lines, areas and topology (i.e. mutual relationships) are also known to the system. If one would like to have a simple map with just the outline of the districts, all three data structures will function. If the map to be drawn is based on the request 'draw only the largest and smallest area', the structure represented in map I will fail. However, the structure represented by map II will also fail when the request would be 'draw only those areas bounded by district C'. The type of request will, for both raster and vector structures, influence the response times.

The nature of the data to be represented will strongly influence the choice between vector or raster data structures. If one is active in the field of utilities the vector approach seems an obvious choice, because of the type of objects one is dealing with (e.g. pipes, networks, etc.). Whenever the organization depends on remote sensing data, a raster data structure is advisable, because it is suitable for both interpreted and non-interpreted data. It should also be realized that a vector data structure is only suitable for data that have been interpreted fully.

4.4.2 Vector approach

The vector structure is one of the oldest structures in use. This is partly because the vector approach is close to the traditional cartographic drafting techniques. Look at an analogue map and it is possible to 'see' lines constructed from nodes and arcs. Another reason is the limited computer technology available at that time. The small computer memories were unable to deal with the vast amount of data involved with raster structures, while vector data structures are relatively small. The basic unit of the vector

data structure is the geographic object. Several kinds of vector data structure exist. For an extensive elaboration on these structures, see Laurini and Thompson (1992). Figure 4.15 illustrates two of these structures.

A non-topological or spaghetti data structure is the most simple type of vector structure. All objects are defined as single items. As can be seen in Figure 4.15(a), the line between points 6 and 7 is defined by a set of coordinates (x_6, y_6) and (x_7, y_7), and is labelled 37. However, the data structure does not include any reference to other objects, such as lines 38 and 34, which are connected to line 37. No reference is made to the areas bounded by line 37. It is similar to map I in Figure 4.14(c), and does not refer to any spatial relationships among the objects defined in the structure. It is unable to answer questions that are not related to drawing its content. Checking its consistency can only be done visually. This approach was introduced at the beginning of the 1970s.

More advanced are those vector data structures that contain topology. Topology defines the mutual relations among spatial objects, and can be used to check consistency among point, line and area objects, or help in finding answers to more complex queries. Topology can also be described as the highest level of generalization possible. Graphically this can result in different images of the same area, as can be seen in Figure 4.15(b). Area C in Figure 4.15(b)I has its 'natural' boundaries. In II and III these boundaries have been strongly generalized, but relations between area C and its neighbours are still valid.

One of the first vector data structures that included topology was DIME. This Dual Independent Map Encoding system was developed by the US Bureau of the Census to deal with census data. An important difference with the spaghetti structure is the incorporation of spatial relationships among the objects registered by the structure. In the 1980s the Bureau of the Census replaced DIME with the more advanced TIGER system (Marx, 1990). More complex and advanced is the georelational data structure. As well as topology, it includes links to a database system containing attribute data. This results in an efficient and flexible structure, and can be found in many of today's GISs, among them Arc/Info. Figure 4.15(c) shows the principles of this approach. The basic unit is the line segment. From each segment the beginning and end points (the 'from' node and the 'to' node respectively) are known. Those two points give direction to the segment, and make it

(a)

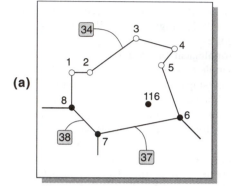

	code	coordinates	attribute
point	116	(x,y)	town
line	34	(x_8,y_8) , (x_1,y_1),........, (x_6,y_6)	boundary
line	37	(x_6,y_6) , (x_7,y_7)	boundary
line	38	(x_7,y_7) , (x_8,y_8)	boundary

(b)

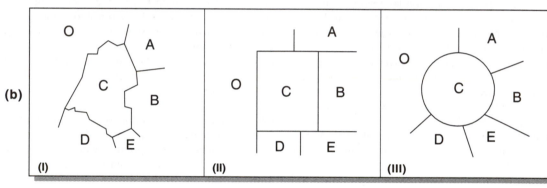

(c)

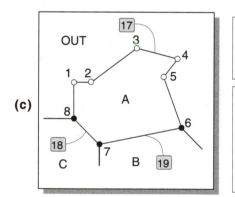

polygon	area	perimeter	attribute-1	attr.
A	783	1621	name	

line	length	from node	to node	polygon left	polygon right	attribute-1	attr.
17	1032	8	6	OUT	A	external boundary	
18	211	8	7	A	C	external boundary	
19	378	6	7	B	A	external boundary	

points	coordinates	attribute-1	attr.
1	x,y	....	....
2	x,y	....	....
...	...	....	....
...	...	....	....
8	x,y	....	....

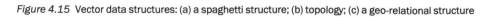

Figure 4.15 Vector data structures: (a) a spaghetti structure; (b) topology; (c) a geo-relational structure

possible to define a left and a right area. In Figure 4.15(c) the line labelled 17 has point 8 as a begin node and point 6 as an end node. The left and right areas are OUT and A respectively. When applicable, the number of points between both nodes is registered. The segment labels can be used as pointers to refer to other tables with information on points, areas or attributes. They function as a link to the database as well. The georelational data structure allows for flexible search operations, for area aggregation, for linking attribute data and for consistency checking.

4.4.3   Raster approach

During the last decade the use of raster data structures has increased. Although it is difficult to code during input, the speed of the scanner, as well as that of output equipment such as laser and electrostatic plotters, offers advantages. Physical storage is simple and stored data are easily accessible. A wealth of processing techniques are available from the remote sensing and image processing disciplines. In the GIS environment this has led to the development of cartographic modelling techniques (Tomlin, 1990).

In the raster approach, spatial units function as basic reference units, instead of as geographic objects (as is the case with the vector data structure). Squares are most common (Samet, 1990) since manipulations with squares are easily performed by the hardware (e.g. pixel on a screen or paper). Figure 4.16(a) shows how a geographic feature is registered in a simple raster structure. The grey squares define Cumbria, and the white squares the area outside Cumbria. Introducing more grey values allows for the registration of more different objects (for instance, different grey values for each of Cumbria's districts). However, it is not possible to register more than one attribute for each square. If there is a need to have more attributes, which occurs especially in a GIS environment, one has to store more raster layers. The size of the squares (the resoluton) will define how well a raster structure can represent geographic reality. A small size will give a better representation, but will result in a very large data set. Sometimes the data source defines the resolution. An example is the resolution of the satellite's scanner, which set condition for data collected by remote sensing techniques.

Another advantage of the square above a triangle or a hexagon is that it can be split into subunits of the same shape and orientation (Figure 4.16b). The quadtree data structure (Samet, 1990) is based on this approach, and is used in GIS packages such as SPANS. Considering the quadtree, the whole map is seen as a square and is subdivided into four smaller squares (see A, B, C and D in Figure 4.16c). Splitting the squares continues until each single square has a homogeneous content. In practice this goes down to six or seven levels. The example in Figure 4.16c is four levels deep. The tree below the map image shows how it is done. Each solid circle represents Cumbria, and each open circle the area outside Cumbria. When no branches leave a circle it is considered homogeneous (black or white). Several variations on this basic approach exist (Samet, 1990).

4.4.4   Hybrid use of the database

Both raster and vector data structures are used via database management systems (Healey, 1991). A database management system often has one of the following structures: hierarchical, network or relational (Figure 4.17). The first has a fixed tree-like structure, and questions can only be asked along the tree's branches, as can be seen in Figure 4.17(a). Here Cumbria is divided into districts, each of which is further subdivided into wards. Because of its fixed structure it has relatively short response times to queries. It is applied mainly in those information systems with management tasks. Here the nature of the questions is known beforehand.

When a more flexible approach is needed during the query process, a network structure can be used. As Figure 4.17(b) demonstrates, this structure is not limited to queries along branches. Database elements of the same level, such as the districts in the figure, can be combined in the query. These databases can be found at utility companies. However, as with hierarchical databases, the type of question is fixed at the moment one defines the database structure.

A relational database structure is even more flexible. It can handle any kind of query, and is very useful in an environment with unpredictable and constantly changing queries. However, it is often slower than the other two database structures. Figure 4.17(c) gives an example. In the GIS database the geometry of Cumbria's districts is stored in a table. The relational database principle allows one to link this information with any other table when a common variable is available. In the figure this is the district's name

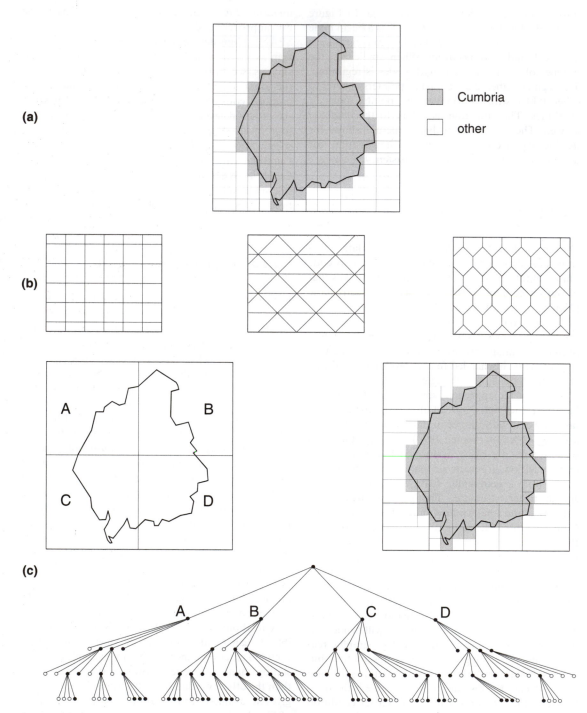

Figure 4.16 Raster data structures: (a) normal; (b) basic raster types; (c) quadtree

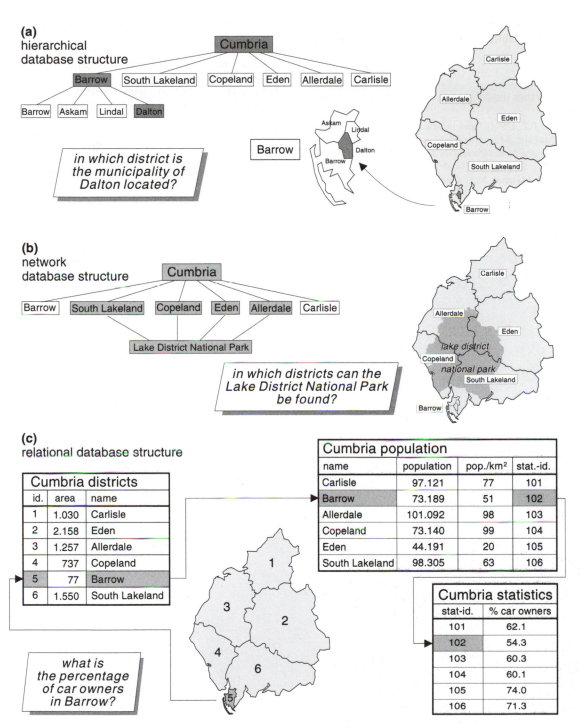

(a)
hierarchical
database structure

*in which district is
the municipality of
Dalton located?*

(b)
network
database structure

*in which districts can the
Lake District National Park
be found?*

(c)
relational database structure

Cumbria districts

id.	area	name
1	1.030	Carlisle
2	2.158	Eden
3	1.257	Allerdale
4	737	Copeland
5	77	Barrow
6	1.550	South Lakeland

Cumbria population

name	population	pop./km²	stat.-id.
Carlisle	97.121	77	101
Barrow	73.189	51	102
Allerdale	101.092	98	103
Copeland	73.140	99	104
Eden	44.191	20	105
South Lakeland	98.305	63	106

Cumbria statistics

stat-id.	% car owners
101	62.1
102	54.3
103	60.3
104	60.1
105	74.0
106	71.3

*what is
the percentage
of car owners
in Barrow?*

Figure 4.17 Database management systems: (a) a hierarchical structure allows for fixed queries along the tree's branches; (b) a network structure offers greater flexibility and allows a combination of items on the same level; (c) relational structures are most flexible, and allow any kind of question on the data stored in related tables

and its statistical identifier. The links between the different tables allow one to ask complex questions.

Further reading

Dale, P.F. (1991) Land information systems. In D.J. Maguire, M.F. Goodchild and D. Rhind (eds) *Geographical Information Systems*. London: Longman.

Healey, R.G. (1991) Database management systems. In D.J. Maguire, M.F. Goodchild and D. Rhind, *Geographical Information Systems*. London: Longman, pp. 251–267.

Laurini, R. and D. Thompson (1992) *Fundamentals of spatial information systems*, the APIC Series no. 37. London: Academic Press.

Peuquet, D.J. (1984) A conceptual framework and comparison of spatial data models. *Cartographica*, **21**(4), 66–113.

Samet, H. (1990) *The design and analysis of spatial data structures*. Reading, MA: Addison-Wesley.

Tomlin, C.D. (1990) *Geographic information systems and cartographic modelling*. Englewood Cliffs: Prentice Hall.

CHAPTER 5

Topography and base maps

5.1 Georeferencing

The locational component gives spatial data their unique character. It distinguishes these data from all other data. Another typical characteristic is that the geographical objects spatial data represent, occur in complex and irregular patterns. This makes it difficult to describe these objects. Try, for instance, to give a description of the basin of the River Zambesi. It should include the shape of the river, its width and its relation with its tributaries, as well as phenomena such as swamps, islands and forests, etc. Also included should be relationships with features such as bridges, dams, other infrastructural artefacts, and administrative boundaries.

Depending on the scale at which this description of the digital landscape model is based, it can be very complex and extensive, but it will probably still be incomplete. Selections have to be made, and terminology used such as 'close to' and 'left of' will not increase the accuracy of the description. What methods are available to indicate location? Each georeferencing method has its own advantages and disadvantages, as Figure 5.1 shows. Names of geographic features (toponyms) can be used. These distinguish one feature from another. However, their use is no guarantee that a unique description will result. 'Springfield' could be in Massachusetts, Illinois, Ohio or Oregon in the United States, or in New Zealand or even in South Africa. A practical approach seems to be the use of address and postcode (or ZIP code). This has an implicit order: 10 Station Road will be close or next to 8 Station Road. However, it only refers to buildings, and does not include natural features like rivers, lakes or mountain ranges. The use of topology is an other option. The

terminology is familiar: Allerdale bounds Eden in the west. However, it does not include any information on real distances (Figure 5.1e). The use of a coordinate system is also quite common. It could be a global coordinate system, where each location is defined by longitude and latitude, or it could be a position in a national coordinate system. National coordinates provide a unique indication of location, but are less natural in use than a system of parallels and meridians. Local coordinate systems for small areas are in use as well. A typical cartographic reference system is the use of map sheets. This method is simple to use, but in practice locations needed are often on two or more sheets. Figure 5.1(f) shows that Keswick can be found on sheet 90 of the Ordnance Survey's 1:50 000 map series.

Geographical information systems always use coordinate systems, often in combination with topology. As explained in Chapter 1, GIS is about data integration and spatial analysis. In order to be able to combine two or more data sets to execute a spatial analysis or a cartographic compilation, it should be possible to reference both sets in a common coordinate system. In practice this is often quite difficult. Different coordinate systems are used (if any), as well as different map scales. This chapter elaborates on the methods and techniques available to deal with these problems. They relate to converting one system to another, the simplification of maps, and the conversion from three-dimensional reality (the globe and local relief) to the two-dimensional plane. Without solving these problems, or knowing their exact nature, spatial analysis or cartographic compilation will not give a valuable result and consequently the maps produced cannot be trusted.

Since the earth is almost a perfect sphere, the elementary globe referencing system is based on sphe-

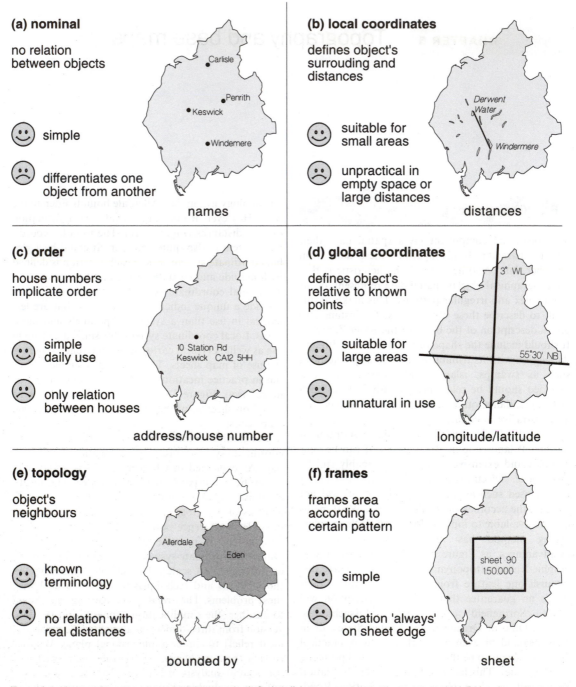

(a) nominal

no relation between objects

☺ simple

☹ differentiates one object from another

names

(b) local coordinates

defines object's surrouding and distances

☺ suitable for small areas

☹ unpractical in empty space or large distances

distances

(c) order

house numbers implicate order

☺ simple daily use

☹ only relation between houses

address/house number

(d) global coordinates

defines object's relative to known points

☺ suitable for large areas

☹ unnatural in use

longitude/latitude

(e) topology

object's neighbours

☺ known terminology

☹ no relation with real distances

bounded by

(f) frames

frames area according to certain pattern

☺ simple

☹ location 'always' on sheet edge

sheet

Figure 5.1 Maps and location: georeferencing methods for localizing geographical objects: (a) nominal; (b) local coordinates; (c) order; (d) global coordinates; (e) topology; (f) frames

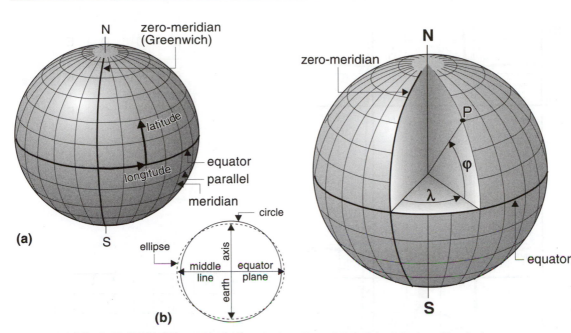

Figure 5.2 Geographical coordinates: (a) longitude and latitude; (b) the earth's ellipsoid

Figure 5.3 Defining latitude and longitude

rical coordinates. This system, called the geographical coordinate system, defines a location by latitude and longitude. Figure 5.2(a) shows this system, where latitude and longitude are measured in degrees, minutes and seconds. The origin of this spherical system is the intersection of the Equator and the Greenwich Prime Meridian. For latitudes, the Equator is defined as 0°, the North Pole as +90°, and the South Pole as −90°. The intersections of all planes of a certain latitude and the Globe are called parallels. All half-circles from the North Pole to the South Pole are called meridians. East of the Greenwich Meridian, which is defined as 0° longitude, they increase up to +180°, and west to −180°. Both parallels and meridians make up the global graticular network. From Figure 5.2(b) it can also be seen that the earth is not a perfect circle. It is flattened at both poles, resulting in the three-dimensional shape of the earth being an ellipsoid rather than a sphere. The ellipsoid is used for calculation purposes to convert the three-dimensional earth to flat paper (see next section).

To define a location, its latitude (φ) and longitude (λ) are measured from the earth's centre to the location on the earth's surface. From Figure 5.3 it can be seen that a point's geographic latitude (φ) is defined by the angle, in a meridian plane, between the

Equator and the line from the earth's centre to this point on the globe. Geographic longitude (λ) is defined by the angle, in the Equator plane, between the Greenwich Meridian and the meridian of the particular point. Accordingly, the location of Helsinki can be defined as 60° North of the Equator and 25° East of Greenwich. To be more precise this would be 60°10′N and 24°58′E. The relation between the geographical coordinates and distances on the earth is determined by the place on earth. Along the Equator, and along all meridians, 1° is 111.11 km in distance, assuming that the circumference of the earth is 40 000 km and has a radius of 6370 km. At 45°N or S, a parallel has a circumference of 28 301 km, resulting in degrees that are 78.6 km in length. At both poles this length is zero (Figure 5.3).

Most national mapping organizations have introduced a national coordinate system for unique and consistent reference and calculation purposes. Figure 5.4 shows the British system. It is called the National Grid system and it adheres to a Cartesian rectangular coordinate system. The map in Figure 5.4(b) shows how Britain is covered by 100 km squares. These squares are a refinement of a 500 km square grid (Figure 5.4a). In relation to the geographical coordinate system, the true origin is found at the crossing of meridian 2°W, and parallel 49°N. The relation

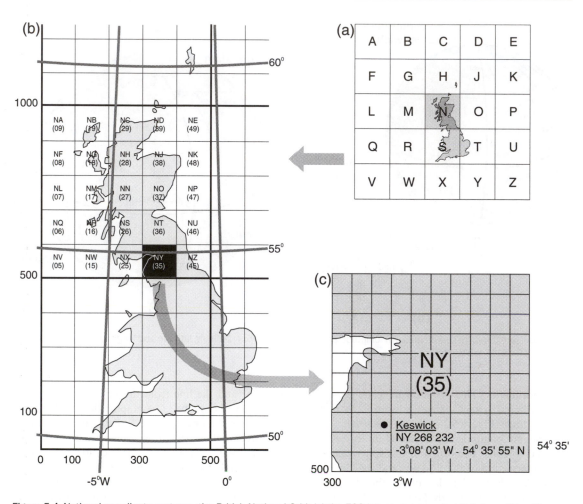

Figure 5.4 National coordinate systems, the British National Grid: (a) the 500-km square covering Britain and its wider surrounding; (b) Britain covered by 100-km squares; (c) defining a coordinate in the National Grid

between both systems is expressed in the figure by the grey graticule lines of the transverse Mercator projection and the black National Grid lines. However, to avoid confusion with negative and positive coordinates when defining a location in the National Grid system, a false origin has been introduced. The true origin has been shifted 100 km north and 400 km west, always resulting in positive coordinates in either x or y direction. To make the system even easier to use, a location is denoted by letters as well. The 500 km square provides the first letter (for Keswick in Cumbria, an N) and the second letter is derived from one of the twenty 100 km squares that make up the 500 km square. For Keswick this is a Y. In Figure 5.4(b) other letters have been added to the 500 km

square N as an example. In this system the 100 km squares can be further refined, as Figure 5.4(c) shows. The location of Keswick can thus be pinpointed by NY268232. This possibility of refinement guarantees that all map scales produced by the Ordnance Survey are covered by the same National Grid.

Figure 5.5 demonstrates the use of a local coordinate system. To position objects found during an archaeological excavation along the River Nile in Egypt, a local Cartesian grid has been created to cover the site. To do so, a base measurement was effectuated along a horizontal railway track between Elkab and Nag'Hilal. Then a triangulation network was set up, based on this baseline. In the actual excavation area a rectangular grid was established, with

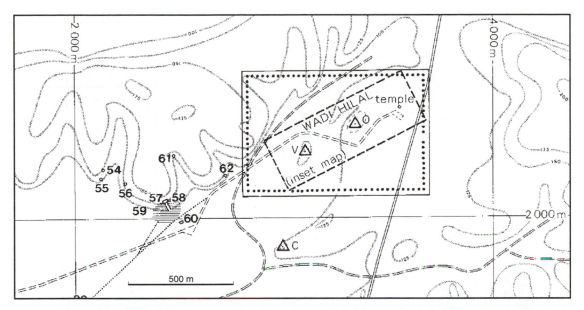

Figure 5.5 Local coordinate systems: the system shown has been established for an archeological excavation along the Nile in Egypt (from Depuydt, 1989)

as its nodes positions determined in this triangulation net. This rectangular grid was densified further locally according to requirements.

5.2   Map projections

Once upon a time cartographic education mainly consisted of training in plotting different projections. Since the introduction of the computer this has no longer been the case, one of the first major contributions of the computer being the ability to plot any area with known coordinates according to any projection. As important as plotting an area according to one of these projections is the ability to spot the nature of projections maps are plotted in, as this knowledge is a prerequisite in dealing with geographical information in GISs. All the spatial operations possible in a GIS are only relevant if the files combined are all based on data in the same reference frames or projections.

Mathematical formulas are used to transform spherical geographical coordinates to the two dimensions of a plane. This transformation process is referred to as map projection. The transformation from the three-dimensional ellipsoid to the two-dimensional plane is not possible without some form of distortion. The distortion affects shapes, distances and directions. Each of the many formulae available (Snyder, 1987, 1989; Snyder and Stewart, 1988; Canters and Declier, 1989; Maling, 1992) will result in different distortions. This determines whether each map projection will be suitable or unsuitable for a certain purpose.

Map projections can be categorized on the basis of the shape of the projection plane, as seen in Figure 5.6. From this perspective, conical, cylindrical and azimuthal projections are distinguished. The point or line where the projection plane touches the ellipsoid is called the point or line of tangency. No distortion is found at this point or along this line. Azimuthal projections have a single point of zero distortion. In the normal aspect, meridians are straight lines, and parallels are concentric circles centring on the Pole. Cylindrical projections have a single line, called the standard line, of no distortion. In the normal aspect this line touches upon the Equator. Both parallels and meridians are straight lines perpendicular to each other. Distortion increases dramatically toward both poles, which are represented by lines.

The conical projection also has a line of zero distortion, as seen in Figure 5.6(a). In its standard aspect the meridians, again, are circular arcs. In

(a)

(b)

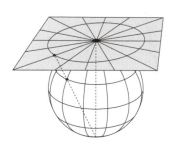

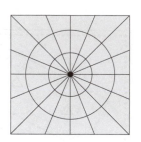

I azimuthal

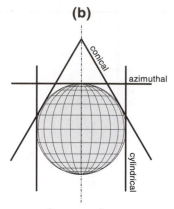

I normal

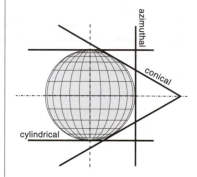

II cylindrical

II transverse

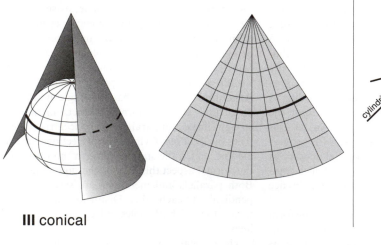

III conical

III oblique

——•—— } point/line of tangency

Figure 5.6 Map projections. (a) Projection planes: I, azimuthal projection plane: II, cylindrical projection plane; III, conical projection plane. (b) Projection aspects: I, normal aspect; II, transverse aspect; III, oblique aspect

Figure 5.6(a) both the cylindrical and conical projections have a single line of tangency (the standard line), resulting in an increased distortion away from these lines. To decrease this effect some projection planes intersect the ellipsoid, resulting in two lines of zero distortion, thus decreasing the total distortion.

The examples in Figure 5.6(a) are all in direct or normal aspect, i.e. the projection form that provides the simplest graticule and calculations. For an azimuthal projection this is the one which touches at the Pole; for conical or cylindrical projections this is the one in which the axis of the cone or cylinder coincides with that of the ellipsoid. Next to the normal aspect this can result in a transverse or oblique aspect (Figure 5.6b). The transverse aspect is characterized by a distortion pattern that, in respect to the normal, is rotated by 90°. The oblique aspect includes all possible cases between the normal and the transverse aspect.

Each projection has its specific patterns of distortion, with the least distortion at the tangential point or line. In order to have as little distortion as possible, it is feasible to change the map projection aspects, and by doing so move these tangential points or lines, so that the area of least distortion overlaps the region to be mapped.

The nature of the distortion pattern is another parameter with which to classify map projections. It is closely linked to the question of which map projection is used. This question will be addressed later in this section. In the transformation from a three-dimensional sphere to two-dimensional plane only a few characteristics can be preserved. These refer to object shapes, to area sizes or to distance between objects. On the basis of the characteristics they retain, map projections are classified as conformal, equal-area or equidistant or as being none of these. If the nature of the mapped phenomena requires that the shapes of the features should be preserved, a conformal map projection has to be chosen. However, preserving shape can only be done for small areas, and it is impossible to maintain it all over the mapped earth. The application of a conformal projection guarantees that local shapes will be retained: circles drawn on a sphere will have circular shapes on the plane as well. However, the size of the circles will be different and may be greatly distorted, depending on the location on the sphere in respect to the point or line of tangency. If one wants to preserve the size of areas, an equal-area projection has to be chosen. When moving a circle over the globe the area covered by the circle will be the same at any location on the globe. Applying an equal-area projection, the resulting area on the plane will cover the same area as on the sphere. However, its shape can be quite different. A circle can be distorted to an ellipse, but still cover the same area as the corresponding circle on the globe. This can be seen in Figure 5.9(b). Figure 5.7(a) shows the globe from two positions, highlighting Greenland and Saudi Arabia, which have about the same size. Figure 5.7(b) shows the effect on both areas of the application of an equal-area map projection, and Figure 5.7(c) shows the effect on both areas of the application of a conformal projection.

Equidistant map projections preserve distances between certain points. This can result in maps in which the distances from one or two points to all other points or certain lines are correct. Projections do exist that do not adhere to any of the three distortion characteristics described, and neither are they related to the cylinder, cone or azimuthal projection approach. To reduce distortion for a specific application they are often modified versions of other projections, while some may be totally different.

Two examples will be elaborated in more detail: the Universal Transverse Mercator (UTM) projection and the Robinson projection. The UTM projection is derived from the transverse Mercator projection, which has characteristics similar to those of the Mercator projection. The original Mercator projection is a conformal cylinder projection where parallels and meridian intersect at 90° angles. The projection is useful for navigation purposes (conformality) or for mapping regions near the Equator. As can be seen in Figure 5.8(a) the areas near the poles are very much distorted. This projection is still inappropriately used for world maps in atlases or on wall charts. Because of its characteristics, it presents a misleading view of the world. The transverse Mercator projection (III in Figure 5.8a) has the same characteristics as the original Mercator. In an adapted version this projection is often used for surveying and mapping applications for relatively small areas. Because distortion near the standard line is limited, more than one standard line was introduced, which resulted in the Universal Transverse Mercator (UTM). For this projection the world is covered by a grid as can be seen from I in Figure 5.8(b). The grid patches or zones have a size of 6° longitude by 8° latitude. The zones are referenced by a letter along the meridians, starting with C at 80°S, and a number along the parallels, starting with 1 at 180°W. One standard line exists for each

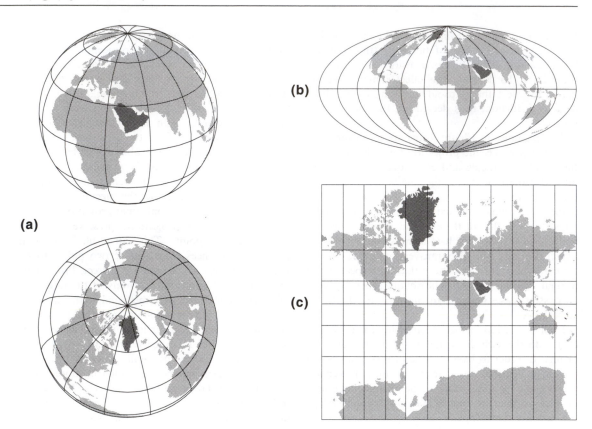

Figure 5.7 Characteristics of map projections: (a) Arabia and Greenland highlighted on the globe; (b) the same areas in an equal-area projection, here Mollweide; (c) the same area in a conformal projection, here Mercator

column of 6°, with 3° on both sides of the standard line. I in Figure 5.8(b) demonstrates this principle. It shows zone 59E and the standard line of 174E. Figure 5.8(c) shows the column between 6° and 0° and locates zone 30U in Western Europe. Each of these zones is further subdivided into several 100 km squares, referred to with letter combinations as can be seen in Figure 5.8(d). The location of Keswick in Cumbria in UTM coordinates is 30U VF913522.

To map the world with as little distortion as possible is quite difficult. Only a few map projections are suitable for world maps: while all projections have disadvantages, some have more disadvantages than others (such as the Mercator projection discussed above). Choosing a map projection for a world map will depend on its intended use. The search will come up with, for instance, a map with conformal characteristics if one wants to visualize dominant wind directions, or with an equal-area projection if one wants to visualize population distribution pat-

terns with the dot map technique. If one is looking for a general purpose map one can go for a compromise between these characteristics. In practice, other factors might influence this decision as well; for instance, the location of the land masses on the earth, or the distribution of the population and its economic activities. Robinson (1974) presented a pseudocylindrical map projection which is neither conformal nor equal-area. It gives the world the 'right' look, but no point is free from distortion. Between 45°N and 45°S distortion is limited, but beyond that it increases quickly towards the poles. Standard lines are 38°N and S (Figure 5.9). Robinson's map projection has been adopted by the National Geographic Society for their world maps.

Which map projection to choose? The path to follow is similar to other decisions to be made in the mapping process. The first step is to define the purpose of the map. Does one want to compare areas, measure distances or display directions correctly? The nature of the distortion is the first guide in the

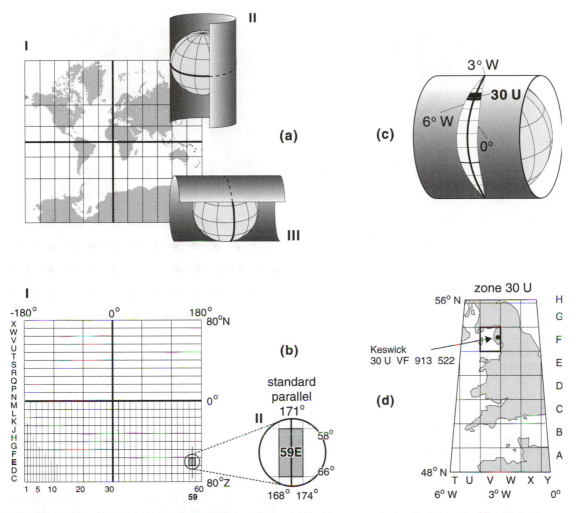

Figure 5.8 The Universe Transverse Mercator (UTM) projection. (a) The Mercator projection (I), its standard line (II), the transverse Mercator projection's standard line (III). (b) The UTM grid (I), zone 59E (II). (c) The location of zone 30U. (d) Zone 30U's 100-km squares and the location of Keswick

selection process. A choice has to be made between retaining conformal, equal-area, equidistant or other projection characteristics. Showing the flow pattern in the atmosphere or oceans requires a conformal projection: comparing vegetation coverage, or just the size of countries, requires an equal-area projection. This type of projection is also needed if one intends to indirectly compare areas in thematic maps, such as dot maps or proportional symbol maps. The second step is to look at the shape, size and location of the geographical area to be mapped. The projection plane and its aspect play a role here. The distortion pattern of the chosen projection

should match the shape of an area as closely as possible. If one looks at Chile, South America, a projection with a N–S standard line would be very suitable since this country has a small E–W extension and a very large N–S extension. A map projection that could be used here is the Transverse Mercator Projection. Russia, on the other hand, has a large E–W extension, but also a relatively large N–S extension. It therefore needs one or two standard line(s) that run E–W. A conic projection would be a useful choice here. Azimuthal projections are very suitable for areas with an equal N–S and E–W extent. Finally, if relevant, the choice should be influenced by the

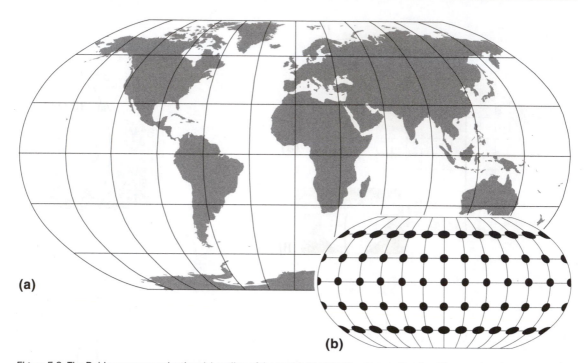

Figure 5.9 The Robinson map projection: (a) outline of the world; (b) distortion: the circles along the equator show correct size. By comparison other circles show the amount of areal distortion at the locations they represent

manner in which the map extent fills the screen or paper. This is especially relevant for atlases. Concluding, one should realize that there are no good or bad projections but, rather, there are good or bad applications of map projections.

5.3   Geometric transformations

In Chapter 2 it was mentioned that it will be unlikely that all the data necessary for spatial analysis or map production will be available in the required format. In Section 5.2 the importance of being aware of the geometrical characteristics of the data prior to their processing and combination in GISs was stressed. This section deals with several solutions that can be applied in order to integrate the geometric component of different spatial data sets. These are the affine and curvilinear transformations. Map projections also belong in this category when one has to convert one projection into another (Figure 5.10). An affine transformation converts all coordinates of a specific data set into the coordinates of another coordinate

system. To apply the transformation, one needs at least three corresponding points in the old and new data sets. It is based on the formulae

$$x' = Ax + By + C$$
$$y' = Dx + Ey + F$$

where x' and y' are the coordinates in the new system. Parameters A, B, C, D, E and F are defined by comparing both points in the two data sets. The transformation is a combination of three basic operations: translation, rotation and scaling (Figure 5.11). The transformation allows for a different scaling along the x-axis in respect to the y-axis.

The curvilinear transformation is applied in the process also known as rubber-sheeting. It is used when two data sets do not geometrically match. However, one should only apply it when there are no alternatives, because it could be a source of error. The starting point in this process is a data set geometrically correct for the purpose it is needed for. The characteristics of the other data set are less well known. This can be due to an unknown map projection or to the fact that a map, used during the digitizing process, has been affected by shrinkage or strains. The relative position of the lesser known map is

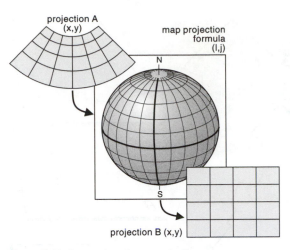

Figure 5.10 Conversion of map projections

reduced to zero. It is this process from which the name rubber-sheeting is derived. From the figure it can be clearly seen that not all points are affected to the same degree, as is the case with the affine transformation. When the lesser known data set is very much distorted it is also possible to execute rubber-sheeting on separate parts of the data set using different parameters. The quality of the result depends on the number of control points and their distribution, but even more on the accuracy of the data set that was assumed to be the better of the two.

A special case of rubber-sheeting is edge matching. Edge matching is applied to correctly connect features that are found at the edges of data sets. This problem is very prominent when data sets that have to be used in a spatial operation are produced by digitizing individual map sheets. Features that cross the map sheet border will almost never connect properly. There are several reasons for this: differences in shrinkage or strains, digitizing inaccuracies, errors in the original mapping process, etc. Figure 5.13 gives an example. It is important to remember that rubber-sheeting can be a source of error. Edge matching should only be applied on those line features which have several points close to the edge, otherwise distortion will occur in other parts of the data set.

defined by clearly identifying corresponding points on both maps. These could be road crossings or characteristic points in a coastline (Figure 5.12). Choosing these points is done with both data sets visible on the screen. The corresponding points will be linked by a vector. The transformation pulls and pushes the lesser known data set until all vectors are

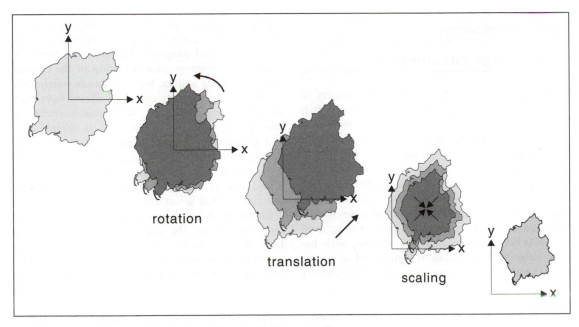

Figure 5.11 Geometric transformations: rotation, translation and scaling

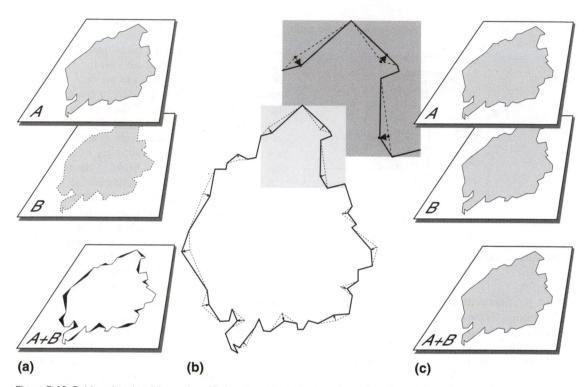

Figure 5.12 Rubber sheeting: (a) map A and B do not match; (b) the transformation of map B based on map A; (c) maps A and B match

(a)   **(b)**   **(c)**

5.4   Generalization

5.4.1   Background and concepts

Each map within a certain scale range requires its own level of detail depending on its purpose. It is evident from Figures 3.3 and 3.4 that large-scale maps usually contain more detail than small-scale maps. However, even at the same scale the level of detail can change. A map for a reference atlas will hold much more detail than a map of the same area in a school atlas (see Plate 12). The process of reducing the amount of detail in a map in a meaningful way is called generalization. The process of generalization is normally executed when the map scale has to be reduced. Map details in Figure 5.14 show why generalization is necessary. The map of Luxemburg in Figure 5.14(a) is drawn at a scale of 1:3 million. Figure 5.14(b) is a photographic reduction of the same map at a scale of 1:12 million. It still contains the same amount of detail as the maps in Figure

5.14(a). The result is a blurred map image. Some text and lines are no longer readable and some have disappeared altogether. Figure 5.14(c), however, is not just a photographic reduction of the original map at scale 1:3 million, but a version in which some symbols have been left out while others have been simplified or emphasized. What has happened is best seen by enlarging this map to its original scale and comparing it with Figure 5.14(a). This also demonstrates that even enlargement is not without implications. The enlarged map detail in Figure 5.14(d) has an unrealistic emptiness and an information density that is too low for its scale.

Especially in a digital environment it is most tempting to use the offered zoom functionality. Unlimited zooming in and zooming out can result in unreliable maps when used to interpret their content. These options should be handled with care, or specific triggers should be built in. An example is the digital *Times World Map and Database* (1994). Figure 5.15(a) shows the coastline of southern South America. Only at the smallest scale will the

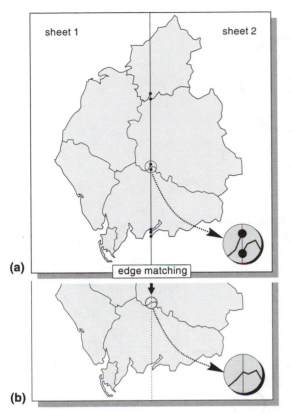

(a)

edge matching

(b)

Figure 5.13 Edge matching: (a) sheets 1 and 2 do not match; (b) after local transformations both sheets do match

coastline be generalized. Figure 5.15(b) zooms in on East Africa. It adds additional map layers with features such as country boundaries, towns, roads and rivers as one zooms in. However, with products like these, the individual layers keep their level of detail throughout the zoom process. Since many GISs have their data organized in layers, this layer-based generalization approach is a standard one in products like these. These products have their role to play, even if, from a cartographic point of view, they are not always correct. In a GIS environment generalization tools should be at hand. If a spatial analysis operation requires data from different sources, these data sets should be of the same level of detail, otherwise it will be (in a logical sense, if not in a technical sense) impossible to use the data in a spatial operation. Results would be unpredictable. GIS digital landscape models contain a selected description of reality as explained in Chapter 1. It is still likely that a further selection may be needed. One reason for this could be that the digital cartographic model requires a much smaller scale than the basic data in the landscape model will allow. Figure 5.16 shows how generalization fits in between the DLM and the DCM.

During generalization one should keep several factors in mind in order to achieve proper results. Most important, as often in cartographic operations, is the

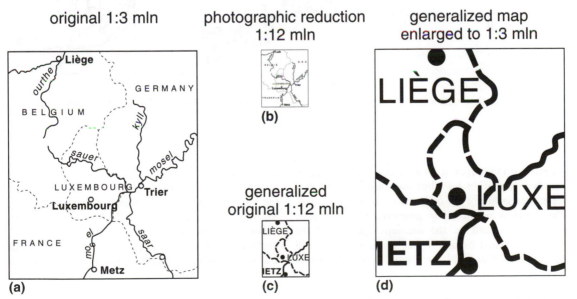

Figure 5.14 Generalization and scale reduction: (a) original map at scale 1:3 million; (b) original map reduced to 1:12 million; (c) generalized original at 1:12 million; (d) generalized map enlarged to 1:3 million

I

II

III

IV

Figure 5.15 Layer based generalization. Examples from the Times World Map and Database (1994). (a) The coastline of southern South America.

purpose of the map and its audience. Generalization entails information loss, but one should try to preserve the essence of the contents of the original map. This implies maintaining geometric and attribute accuracy, as well as the aesthetic quality of the map. The visual hierarchy (see Section 6.3) should be maintained as well; for instance, prominent features in the original map should remain prominent in the generalized result. Depending on the audience, the results may be different, as in the example of the reference atlas and the school atlas discussed

above. Another important factor is the magnitude of the scale reduction. It is obvious that the larger the reduction, the more radically the generalization will affect the original data. Technical and human factors also influence the generalization process. Technical factors include the size and resolution of a monitor screen. In a GIS environment, considering the computational elements is equally important. Which algorithm is most cost-effective and will result in maximum data reduction and minimum storage capacity? These factors interrelate with the human

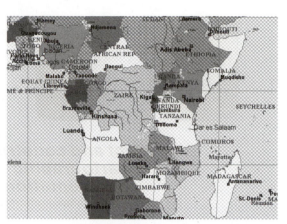

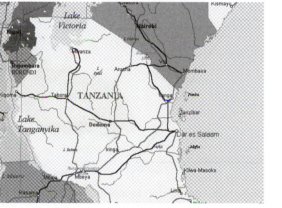

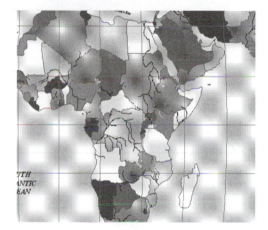

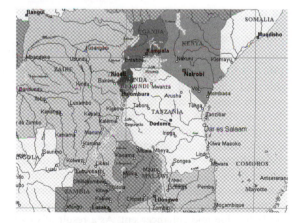

Figure 5.15 (b) Zooming in on eastern Africa

factors. The discriminating capacity of the human eye is limited. Finally, one should always consider the nature of the map contents. Does one deal with quantitative or qualitative information? An answer

to this question defines the setting in which a cartographer or software can execute the generalization process. A qualitative map content requires a different approach from a quantitative content. The for-

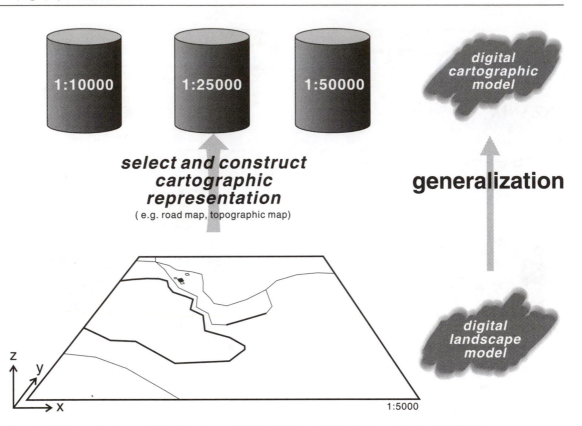

Figure 5.16 Generalization and Digital Landscape Models (DLMs) and Digital Cartographic Models (DCMs)

mer will require a more extensive knowledge of the mapped features when compared to the latter. This split in the nature of the map contents results in two classes of generalization: graphic and conceptual generalization.

Even if everybody took into consideration the factors mentioned above, generalization results could differ from cartographer to cartographer and from algorithm to algorithm. Generalization is and will always be a subjective activity. It is difficult to set up fixed rules, although experiments attempting to do so are manifold. The subjectivity is demonstrated in Figure 5.17. Details of the Nile Delta from different atlases at the same scale are shown. It can be seen that each atlas has chosen its own river branches, with their own level of simplification. Part of the subjectivity has shifted from the cartographer to the programmer who created the algorithm, but the algorithm's parameters are still their user's choice, as is the point where these algorithms are applied first.

5.4.2   Graphic and conceptual generalization

Two types of generalization are distinguished: graphic and conceptual generalization. The difference between them is related to the methods involved in the generalization process. Graphic generalization is characterized by simplification, enlargement, displacement, merging and selection. None of these processes affects the symbology. Dots stay dots, dashes remain dashes, and patches stay patches. Conceptual generalization is also characterized by the processes of merging and selection, and in addition comprises symbolization and enhancement. As a result of these actions, the symbology in the map may change. Another difference is that the processes linked to graphic generalization mostly deal with the geometric component of spatial data, while those processes linked to conceptual generalization mainly affect the attribute component. The difference between both types of generalization is illustrated in Figure

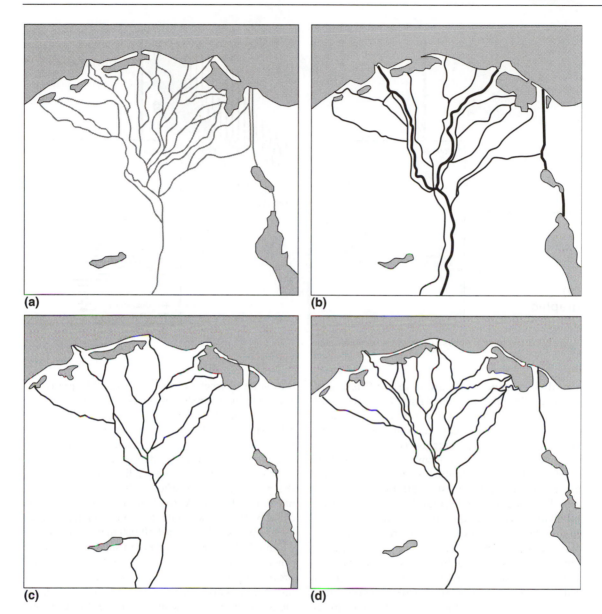

Figure 5.17 Samples of subjectivity in generalization: The Nile Delta at 1:5 million from (a) the *Atlas Jeune Afrique*; (b) the *Alexander Weltatlas*; (c) the *Atlas of Africa*; (d) the *Times World Atlas* (after Pillewizer and Töpfer, 1964)

5.18, showing the process of merging. Figure 5.18(a) shows a forest map of the Netherlands. On top of it, the map with the generalization result, but before scale reduction, is shown. Small individual forest areas in the original map have been merged with those nearby. The maps in Figure 5.18(b) show the geology of the Netherlands. The original map shows the surfaces of the Holocene, Pleistocene and other

periods. In the map on top several periods have been grouped together. In the example of the forests, cartographic common sense could be used to merge the individual patches. However, with a geological map just grouping what is close is not enough. An understanding of the geological timetable and classification system is required. For instance, Holocene and Pleistocene can be grouped together, since in a geo-

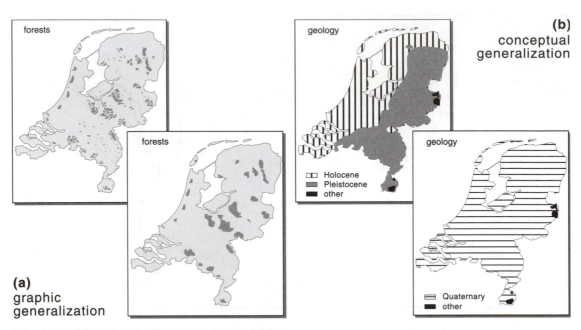

Figure 5.18 (a) Graphic versus (b) conceptual generalization

logical classification both are from the Quaternary. Grouping the small patches classified as 'other' with Pleistocene is not allowed in a geological context. In other words, conceptual generalization requires knowledge of the map contents. To generalize these maps one depends on principles of the discipline involved. Their classification system changes, and this results in a different structure of the legend as well. It should also be realized that, though one can subdivide generalization into several sets of processes, these processes usually interrelate. It is not only simplification or only replacement. Often one process is necessary as a direct result of another process. For instance, when a road has to be enlarged in order to remain visible after scale reduction, then several houses along this road will need to be replaced, otherwise the road symbol will cover the houses.

Figure 5.19 illustrates the processes involved in graphic generalization: simplification, enlargement, displacement, merging and selection. For each of these actions, three illustrations are given: the original map detail, a generalized map detail before scale reduction, and the generalized map detail after scale reduction. Simplification, sometimes called smoothing, should reduce the complexity of the map. The illustration in Figure 5.19(a) shows a river that has a very sinuous nature with many bends. After general-

ization the character of the river should be preserved: a meandering river should still be recognized as such. Enlargement (Figure 5.19b) is sometimes needed, otherwise symbols would disappear, or would no longer be legible after scale reduction. This action affects roads. To keep a road symbol legible it has to be enlarged. It should be realized that the road in the generalized map, after scale reduction, will be much too broad. A road at a scale of 1:10 000 could be 10 m wide, while the same road, represented by a similar symbol, on a 1:50 000 map would have a width of 50 m. Displacement is usually the result of other generalization procedures. It is also a critical procedure since one should take care that, for instance, a symbol representing a house is not placed along the wrong line symbol. Figure 5.19(c) shows the need to displace a house because of the enlargement of a road symbol. In Figure 5.19(d) some individual houses are merged to form a built-up area. Selection, as demonstrated in Figure 5.19(e), is the process of randomly selecting symbols from a set of identical symbols, of which the rest are omitted in the resultant map. It is important to note that omitting symbols should not disturb the overall impression of the phenomenon's distribution. Selection is necessary, otherwise the map image would become too cluttered. The illustration shows how some islands along the coast are left out.

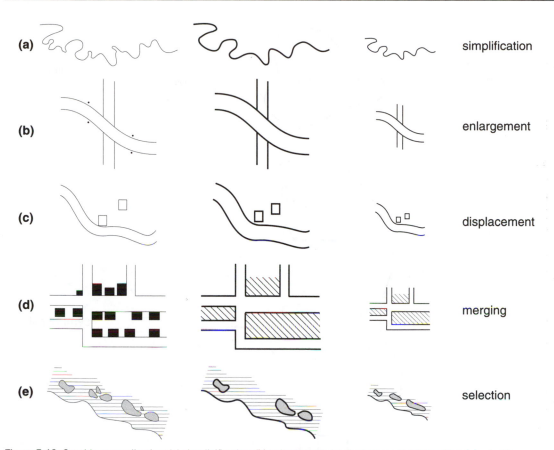

(a)   simplification

(b)   enlargement

(c)   displacement

(d)   merging

(e)   selection

Figure 5.19 Graphic generalization: (a) simplicification; (b) enlargement; (c) displacement; (d) merging; (e) selection

The procedures involved in conceptual generaliza-tion are explained in Figure 5.20. They are merging, selection, symbolization and enhancement/exaggera-tion. Figure 5.20(a) shows that merging of symbols cannot be done without expertise, since it has conse-quences for the legend as well. Some symbols will disappear from the legend, while a small number of new units might appear. Selection in the context of conceptual generalization requires knowledge of the mapped phenomena. In the example of Figure 5.20(b), a lithographic map has symbols for marl, chalk and basalt. Although small in areal extent, the basalt is so characteristic that to leave it out would be to destroy the character of this volcanic island.

Symbolization denotes that the relation between the symbol and the space it represents changes. Dots (e.g. a group of oil rigs) will change into a single area symbol (e.g. an oil field). The moment of change depends on the original scale and the scale after

reduction. Generalization can also result in a map where some symbols attract too much or not enough attention. These symbols have to be enlarged or reduced in size. A main road through a village could become insignificant after scale reduction. So it has to be enhanced by using thicker lines to attract attention proportional to its importance in the map image. Figure 5.21 illustrates some of the generaliza-tion procedures executed when generalizing topo-graphic maps from a scale of 1:25 000 to 1:50 000. See also Plate 12 and Figure 5.34(b).

5.4.3 Generalization processes and tools

The previous section explained the principles of gen-eralization. How does it work in practice? Several cartographers have struggled with this question and suggested conceptual models in an attempt to solve the generalization problem. Among them are

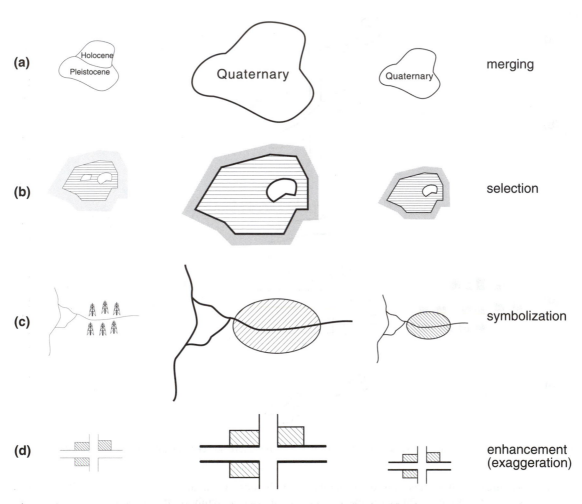

Figure 5.20 Conceptual generalization: (a) merging; (b) selection; (c) symbolization; (d) enhancement

Morrison (1974), Brassel and Weibel (1988) and McMaster and Shea (1992). The latter model, in a slightly adapted version, is used to explain the practice of generalization and the tools available. McMaster and Shea developed their method explicitly for digital cartography. Figure 5.22 summarizes this approach. They decomposed the generalization process into three tasks, translated into the questions of why, when and how to generalize? ˜Why generalize' has been discussed at the beginning of the background section.

'When?' is related to a cartometric evaluation of the original map data in relation to the generalized map. A large reduction factor will cause legibility problems related to graphic congestion and the coalescence of graphic symbols. These problems can be

avoided by the application of, for instance, threshold values or minimum distances between graphical objects. Whenever objects come too close to each other or whenever a graphical density (number of objects within a 10 × 10 cm square on the map, on paper or on the screen) is surmounted, generalization algorithms can be started up. McMaster and Shea describe this as the transformation control. It consists of the selection of suitable algorithms and of the suitable parameter values to apply to them.

'How?' is related to the tools that can transform the geometric and attribute component of the spatial data, to generalize effectively the original map data. Here the tools, or rather the algorithms, that execute the actions of the graphic and conceptual generalization, are discussed. The generalization of attribute

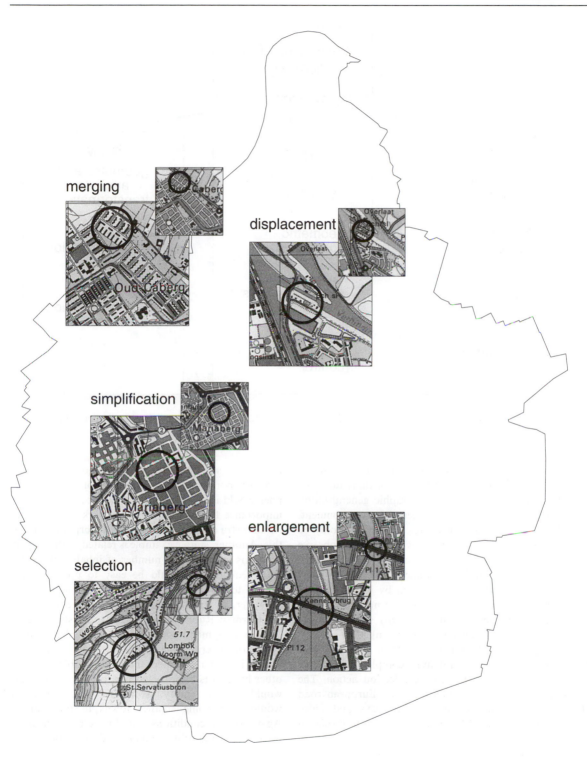

merging

displacement

simplification

enlargement

selection

Figure 5.21 Examples of graphic generalization from topographic maps: Maastricht 1:25 000, sheet 69B and 1:50 000, sheet 69W (Courtesy Topografische Dienst Nederland). See also Plate 13 and Figure 5.34(b).

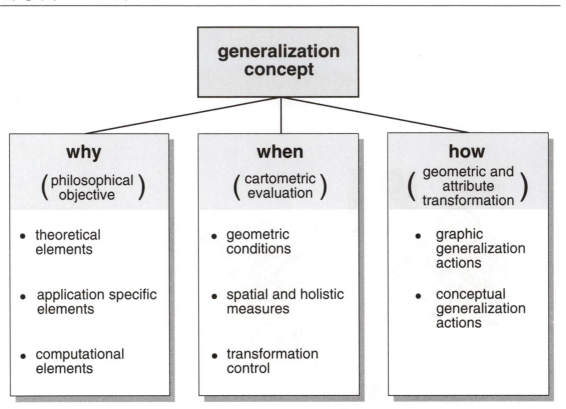

Figure 5.22 A conceptual model of generalization (after McMaster and Shea, 1992), dealing with the questions why, when and how to generalize a data set

data (classification and symbolization) will be discussed in Chapter 7. Many algorithms do exist. Most of them are related to graphic generalization actions, in a vector or raster data environment. This is not strange, since these actions can only be translated into an algorithm when one can observe regularities in them. For most aspects of graphic generalization this can be effectuated. An extensive overview of these algorithms is given by McMaster and Shea (1992). However, it should be noted that most of the algorithms available can only solve relatively simple isolated problems that are linked to a single basic graphical element such as lines. Some of these algorithms can still be relatively complex. Let us look at an example related to the selection action. The starting point is a database of the European road network, which contains all motorways, and major and minor roads. When part of it has to be displayed at a reduced scale, many roads would have to disappear. It would be simple to suggest to select just the motorways. However, this would result in a European network full of gaps. Often, near frontiers,

motorways stop and change into major roads which cross the border. In a manual situation the cartographer would incorporate these roads because of their importance to the total road network. In the algorithms extra conditions have to be incorporated to do this. Another, simple example is related to the display of towns based on the number of inhabitants. After scale reduction it would be easy to preserve only those towns with over 100 000 inhabitants. However, in very densely populated areas such as the German Ruhr area the map would still be cluttered with symbols and text. Cartographers would apply other criteria as well to omit some more towns in order to keep the map legible. On the other hand, lots of areas with only a few inhabitants would remain empty. Normally a cartographer would keep some smaller towns in these areas. Again, special conditions would have to be incorporated in the algorithm to make it do just this.

The generalization algorithms available in most GIS packages are related to line simplification. Figure 5.23 shows the result of two algorithms. In

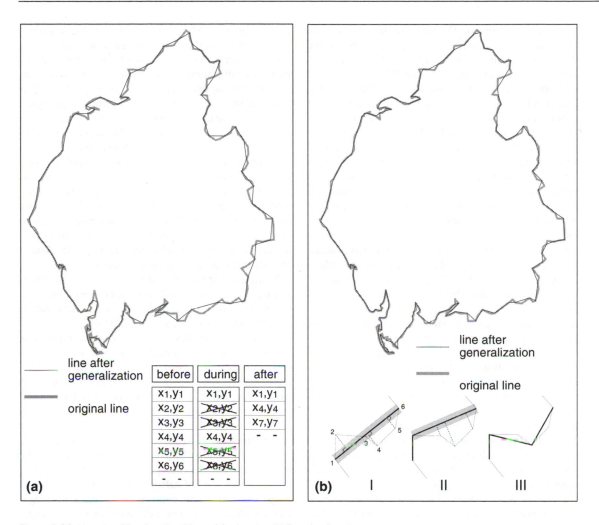

before		during		after	
x₁,y₁		x₁,y₁		x₁,y₁	
x₂,y₂		~~x₂,y₂~~		x₄,y₄	
x₃,y₃		~~x₃,y₃~~		x₇,y₇	
x₄,y₄		x₄,y₄		- -	
x₅,y₅		~~x₅,y₅~~			
x₆,y₆		~~x₆,y₆~~			
- -		- -			

(a)

(b)   I   II   III

Figure 5.23 Line simplification algorithms: (a) *n*th point; (b) Douglas-Peucker

Figure 5.23(a) the nth-point algorithm, which retains every n^{th} point, is explained. It follows a simple approach, and does not consider any relations between neighbouring points. The grey line represents the original data, and the black line the generalized data. In the example n is set to three. During execution, only each third point is retained as can be seen in the table below the map. The user can define n, and with it the magnitude of generalization. The result will depend on the density and homogeneity of the points along the line. A large value for n will result in a strongly generalized line, and very likely in the loss of original line characteristics. An algorithm that considers the line characteristics is the Douglas–Peucker algorithm. It has been accepted as

one of the best generalization algorithms available, and has been implemented in many GIS packages. The algorithm considers the whole line during the generalization process and eliminates points in an iterative process. It is efficient but relatively slow in processing (Douglas and Peucker, 1973). An example is shown in Figure 5.23(b). Again, the grey line represents the original data and the black line the data after generalization. The algorithm works with a baseline and a tolerance zone. The baseline connects the beginning and end points of a line (here 1 and 6). The first is called the anchor point and the second, the floater. The tolerance zone, defined by the user, should guarantee that the line characteristics are preserved. The smaller the zone, the better the characteristics are preserved. The distance perpendicular to the baseline is calculated for all points between the floater and anchor. The point with the largest distance (point 2 in Figure 5.23(b)II) will be saved, and will function as a new floater point. For the remainder of the line this process is repeated. All points that, at a certain moment during execution of the algorithm, fall within the tolerance zone are considered not relevant, and are left out in the final result (points 3 and 5 in Figure 5.23(b)). This figure illustrates a very simple case, and in practice many iterations have to be executed. The result depends on the number of points in the original line, their distribution and density, as well as the width of the tolerance zone.

Generalization in the raster domain means generalization of the attribute data, since the algorithms available simplify the image. The most simple approach is to define the dominant grid cell or pixel attribute for a region and give this value to all pixels in the region. This approach does not preserve the character of the mapped phenomena, and more clever algorithms have been developed. Most of the raster generalization tools originate from image-processing disciplines. Often the algorithms work with some kind of filter matrix, which are moved over the whole raster data set. Figure 5.24 shows how these filters operate. The upper filter in Figure 5.24II has a low value for the central pixel, which gives the surrounding pixels more weight in the process. The lower filter which will have a less generalizing effect on the final data set because the value of the central pixel is high. The result of the generalization process is given in Figure 5.24(a)III, while in 5.24(b) the effect of both filters on a single pixel is shown. The size of the filter also influences the result.

It has not been possible to translate conceptual generalization processes into algorithms with any great success. The reason is that such processes require knowledge of the mapped theme. However, the potential of knowledge-based systems is likely to change this. Several experiments with knowledge-based generalization systems have been performed (Buttenfield and McMaster 1991; Müller, 1991). Müller tried to apply a knowledge-based system to the generalization of topographic maps in the scale range between 1:25 000 and 1:250 000. Figure 5.25 shows how the graphic and conceptual generalization actions fit in here. Topographic surveys have clear generalization rules, otherwise they would not be able to produce uniform map series that cover a whole country. Müller notes that linear elements tend to dominate the map image more, the further the scale decreases. It has also been noted that the number of object classes decreases as well, and roads become disproportionate in size in respect to other line symbols. Of all geometric relations, topological relations remain the longest whereas shapes, distances, etc., are affected. In the experiment conducted, rules and observations like the one mentioned in Müller's system were applied. This was done under the assumption that the data would be available in raster or vector format, and that the algorithms needed would be available. While going through such a scenario one still finds many unsolved problems. Most rules point out what should not be done, but do not say what should be done. And if they do, they do not say how to do it.

Many problems still exist and several potential solutions are only available in experimental environments. In 1994 Intergraph launched a generalization module (Map Generalizer) in its MGE GIS environment, which provides the user with an interactive generalization environment. The user can select a specific area or a specific data layer, decide which algorithm to use (e.g. for line generalization), and can subsequently set parameters. Results can be directly viewed on the screen, and when results are not as expected, the user can change the algorithm or its parameters. It is also possible to use different parameters or algorithms for different parts of the study area. The operator, however, should still possess a certain level of cartographic skills to be able to judge the results. In combination with knowledge-based systems, this approach looks very promising.

5.5   Relief

5.5.1   Introduction

In mapping the terrain the cartographer has to deal with relief data. Again, there is a struggle to transform three-dimensional reality onto the two-dimensional plane of the screen or paper. Relief display should result in a map that provides a geometrically accurate view of the terrain and its shapes (morphology). The method of relief display depends on the purpose of the map. A map intended to give a global impression of terrain requires a different approach as compared to a map from which one

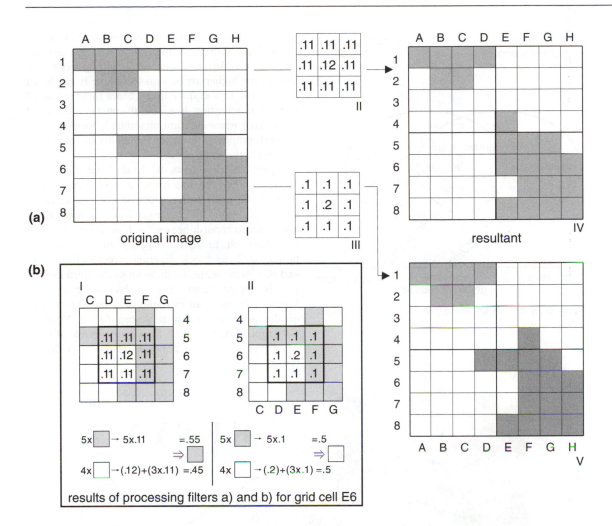

Figure 5.24 Area simplification algorithms: (a) the effect of two different filters; (b) the effect of the two filters on a single pixel

wants to determine heights within accuracy limits of 10 cm. A map to be used for building a dam will contain different height information from a tourist map for a skiing area. It is possible to represent terrain in absolute or relative terms. In Figure 5.26 both methods are shown. Absolute heights can be displayed by contour lines or height points, as shown by the graphic representation of the northeastern side of Mount Kilimanjaro in the figure. Their value is determined above or below a reference plane. Relative height indicates whether a certain location is higher, equal or lower than other locations, which can be effectuated by layer tints, sometimes enhanced with hill-shading, as is shown for the western side of the Kilimanjaro (Figure 5.26).

In mapping relief, cartographers have employed many graphical techniques. The oldest relief maps portray the terrain with simple symbols. Mountains were drawn in aspect or sketched. Relief mapping further developed via hachuring without three-dimensional stimuli, to a systematic hachuring, later followed by hill-shading to enhance the relief impression. Figure 5.27(a) shows hill-shading by the Swiss cartographer Imhof. He introduced the *luftperspektivische Geländedarstellung* to represent relief (Imhof, 1982). It is based on the natural effects of atmospheric colours in the mountains, and is seen as one of the best methods of indicating map relief on a geometrically accurate orthogonal map, while still preserving the third dimension. The shading is

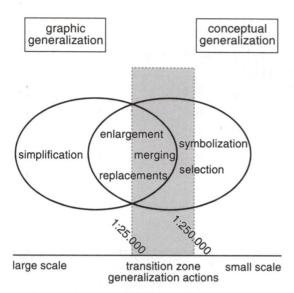

Figure 5.25 Knowledge-based generalization (after Müller, 1991)

printed in tints of violet, and a yellowish tint is added to the slopes facing a fictitious light source. The relief impression is further enhanced by adding more contrast in the higher areas. This technique is employed in the Swiss topographic maps. Since this method of relief representation is a skilful and (as with most other relief representation methods) laborious activity, cartographers have tried to employ the computer to achieve the same effect.

Figure 5.27(b) shows a detail from the *Grote Bosatlas* (1995). In this map, layer tint mapping is applied. All areas between certain contour lines get a specific colour, which ranges from dark green for low areas to reddish-brown for high areas. A different approach to relief mapping entails representations such as block diagrams, perspective views and panorama maps. In these maps the third dimension is not projected on an orthogonal base map. Figure 5.28 gives an example of a panorama map of the Atlantic Ocean produced for a National Geographic map by H.C. Berann. A review of most relief representation techniques is given by Imhof (1982).

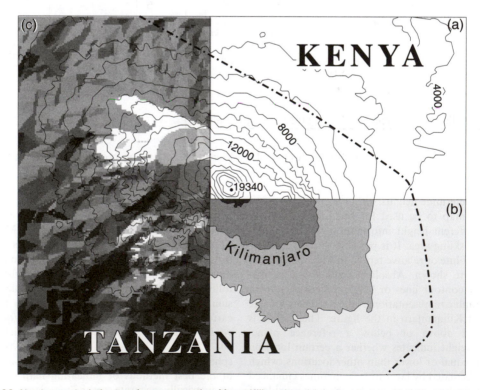

Figure 5.26 Absolute and relative terrain representation, Mount Kilimanjaro: (a) contour lines and height points (absolute); (b) layer tints (relative); (c) contour lines and hill shading (combination)

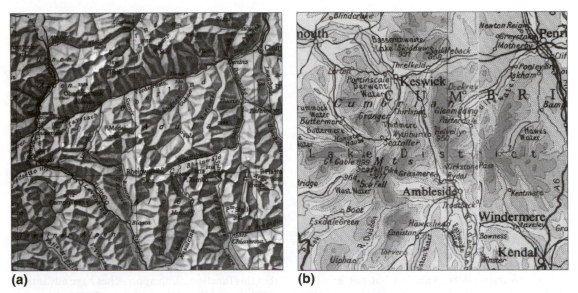

Figure 5.27 (a) Hill shading (detail from *Schweizer Weltatlas*, Courtesy Kanton Zürich) versus (b) layer tint mapping (detail from the relief map of the Netherlands, *Grote Bosatlas*, Courtesy Wolters-Noordhoff)

5.5.2 Digital terrain models

To create relief displays by computer one needs data to start from. The key phrase here is digital terrain model (DTM). DTM is usually defined as a numerical representation of terrain characteristics. When dealing with the altimetric aspects only, they are often called digital elevation models (DEMs). Both DTMs and DEMs can be seen as digital landscape models (DLMs; see Section 1.2) that can be manipulated in a GIS for surface operations, or to create digital cartographic models. There are several applications that require a more general approach to digital terrain modelling allowing for the incorporation of three-dimensional spatial objects that are application-dependent and may be called topographical, geographical, geological, etc. Examples related to geosciences are given by Raper (1989). At the same time, the functionality of existing GIS software packages is being expanded to combine relief data with planimetric data of topographical and thematic coverages. To cover all applications the definition should be expanded: a digital terrain model is a digital three-dimensional representation of the terrain surface and of selected zero-, one-, two- and three-dimensional spatial objects that are related to this surface. The use of three-dimensional maps can be very effective in explaining spatial relationships. For

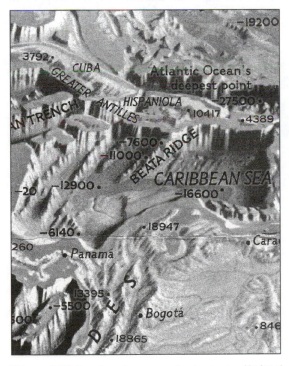

Figure 5.28 Terrain shapes and structure: *National Geographic's* map of the ocean floors (Courtesy National Geographic Society)

instance, when mapping the earth's surface, digital terrain models can given an explanatory insight into its forms. Using the possibilities to look at the terrain in an interactive environment, by changing view angle and azimuth, the sometimes difficult interpretation of the contour line pattern of a topographic map or chart can be avoided. The height information can be combined with, for instance, land use information.

Digital terrain models have found a wide range of applications. They are used in civil engineering for the determination of earthwork cut and fill volumes, landscaping, and to create a visual impression of the environmental impact of civil engineering projects. In topographic mapping they are used to visualize terrain forms. In geological and geophysical mapping they visualize surface and underground structures. They are also in use in navigation simulation, for instance to train pilots. And, last but not least, the military applications have to be mentioned. Here digital terrain models provide information on visibility from a specific point, while slope information is used to plan the most suitable route, and some missile guidance systems use digital terrain model information for navigation.

In order to collect data for digital terrain model building, the same data collection techniques that were discussed in Chapter 2 are used. Existing maps that contain contour lines and spot heights can be digitized. One should note that the quality of the DTM can never surpass that of the map the data were derived from. Because contour lines are often a product of interpolation, the quality of DTMs based on digitized data is often less than those collected by photogrammetric or surveying techniques. This last technique provides the most accurate data, but is only suitable for relatively small areas. The result acquired by photogrammetric techniques depends on the photo scale. Next to the technique, the data gathering method applied will influence the usability of the DTM. If one applies selective sampling, all characteristic points in the terrain can be incorporated in the DTM. Another method applied is systematic sampling. Here data are sampled at regular distances. This approach provides no guarantee that characteristic points such as the highest and lowest point can be incorporated, since they can be located between the sampled points. Selection of a method will depend on the purpose of the DTM (measurements or presentation), the nature of the terrain (roughness and accessibility), and the available hardware and software.

To structure the DTM data in a digital landscape model one can choose between a grid approach and a triangulation approach. The first results in a regular network of points covering the study area. Height values are determined at these points. Figure 5.29 shows an example. In the figure the height values are based on original height measurements as displayed in Figure 5.29(a). The distance between the points, in relation to the distribution of the original data, as well as the local or overall interpolation method applied, defines the model's accuracy. An example of local interpolation is illustrated in Figure 5.29(b). For each point value to be determined, the heights of the six closest original data points, including a distance weight factor are used. An example of an overall approach is the calculation of a polynomial function that includes all original points. The grid point height values can be derived from this function. Both approaches have advantages and disadvantages in respect to the result and to processing time. Many organizations selling DTM data, such as the British Ordnance Survey and the United States Geological Survey, use the grid approach. In Figure 5.29(c) a triangular network is drawn. An important characteristic of this approach is the fact that each of the original data points is incorporated in the model. This offers the opportunity to consider local relief characteristics. The network in Figure 5.29(e) is the result of the Delaunay triangulation algorithm. It results in a triangular irregular network (TIN). This algorithm is incorporated in most GIS packages (often in a slightly adapted version). Characteristic of a Delaunay triangle is that it has edges with the shortest possible length, and the angle between two edges is as large as possible.

In a GIS environment the DTMs are used to execute surface analysis. The basic spatial units of a DTM, grid squares or triangles, have an important function. Two of their attributes, slope and aspect, play a prominent role in a calculation related to the surface analysis (Figure 5.29f and g). Slope is defined as the amount of change in height over a fixed distance, expressed as a percentage or in degrees. Aspect is the orientation of the spatial unit in respect to the north, and expressed in degrees.

5.5.3 Terrain visualization

Questions one is likely to ask when executing a terrain surface analysis range from simple queries such

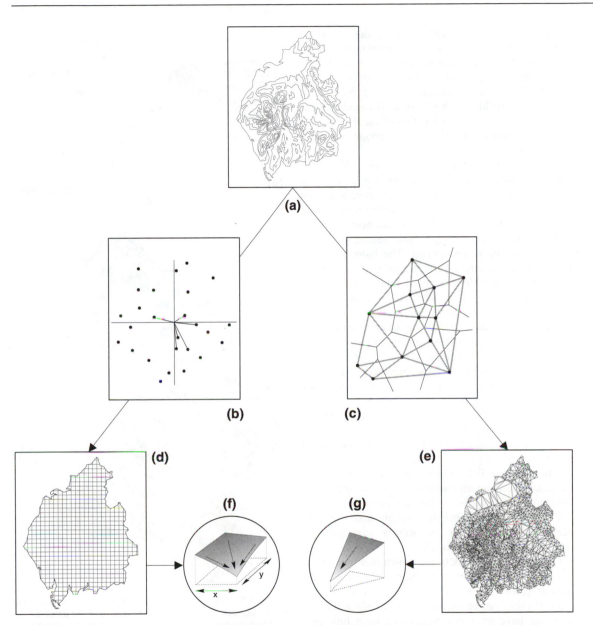

Figure 5.29 Data representation from (a) contour lines to (d) a regular grid and (e) triangular irregular network. The path towards the grid applies an interpolation algorithm (b), and the path to the network applies a triangulation algorithm (c)

as 'What is the height at this location?', and 'What areas can be found between 200 m and 400 m?', to more complex queries such as 'Which terrain is visible from this point?', or 'How much earth has to be moved if . . . ?' Most answers can be given in map form.

The shaded relief map is DTM's most prominent derived product. Shaded relief maps used to be pro-

ducts that could only be created by a few very skilled cartographers. Today this hill-shading technique is available to every GIS user. However, it should be mentioned that, as with generalization algorithms, the users can still make many mistakes, since they have to set lots of parameters. In traditional cartography the sun is put at a certain angle in the north-west to create a shaded relief map that is perceived correctly. Having the light source in the south for instance (which after all is quite natural in the northern hemisphere) would result in a map on which people would see mountains in valleys and the other way round. This relief inversion is caused by stimuli received by the human brain. In a GIS environment the user can freely set the location of the light source and so can create maps that give a wrong impression of the terrain.

Figure 5.30(a) explains how a shaded relief map can be created from a DTM. In the example a TIN-based DTM is used. For each triangle a normal (the vector perpendicular to the triangle plane) can be calculated. Next, the angle between the normal vector and a vector representing the light source in respect to the observer is determined. The light source is placed in the northwest at an angle of 45° to the plane. Working with normalized data, this will result in a value of between one and zero for each triangle. The next step is to create a grey scale, and to define the lightest tint as zero and the darkest tint as one. Now it is possible to match triangle values with a tint from the grey scale. The approach described here is the most simple solution. More advanced algorithms can consider special cases as well. The result of the algorithm is shown in Figure 5.30(b). Here some of the original triangles can still be recognized. This is due to the density and distribution of the original height data as displayed in Figure 5.29. The option to change the algorithm's parameters, such as the position of the light source, is sometimes needed because of terrain characteristics. An example is a mountain ridge with 45° slopes, which ranges NE–SW. Without interference the resultant map would be black and white only – something cartographers would try to avoid if possible.

Other derived products are shown in Figure 5.31. On the left of this figure a profile based on the DTM data is presented, while on the right is a visibility map. Profiles can be used to see what effect the terrain will have on a new high-speed train link, and give insight as to where earthworks have to be excavated. The dark area in Figure 5.31(b) shows the terrain visible from the selected point, looking northwards. The dashed lines indicate the view angle. One can also set the maximum viewing distance. In the examples it was assumed that weather conditions would allow for 30 km visibility. More complex, but also possible, is the calculation of the terrain visible from two different points, or the calculation of the terrain from where one could see the location of the observer. As well as having several military

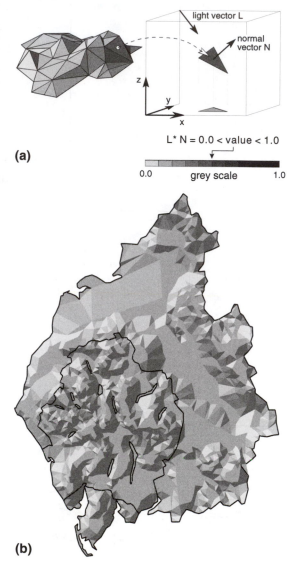

(a)

(b)

Figure 5.30 Principles of hill shading: (a) the grey value of a terrain patch is determined by the values of the patch's normal and the position of the light source relative to the patch; (b) an example of a shaded relief map

applications, this type of calculation for visibility maps is used in environmental impact accessment, for instance in order to determine where to put high rise building without visually disturbing the countryside. Most digital cartographic models from which the figures in this section are derived, neglect the third dimension, although it is available in the digital landscape model. As the panorama maps in

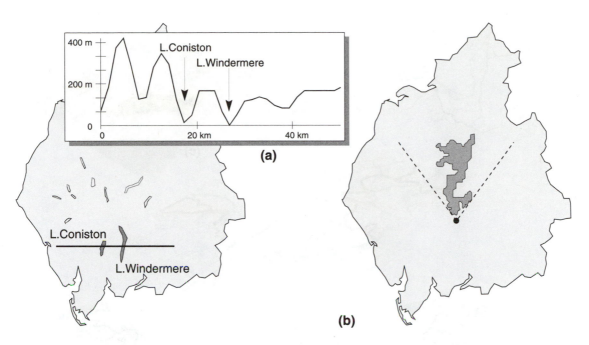

Figure 5.31 Digital terrain models and derived products: (a) profiles; (b) visibility maps

Figure 5.32 demonstrate, it is possible to preserve the third dimension in the image. The main map in the figure shows the terrain with other data, such as hydrography, vegetation and roads draped over it. The inset in the upper left of the figure shows a detail of data in two dimensions, the insets at the bottom show the same terrain from different viewpoints, and the upper right inset shows a fully shaded perspective map with terrain features draped over.

In order to be able to create maps such as those in Figure 5.32, as well as any other three-dimensional maps, the nature of the programme functionality available is very important. Figure 5.33 shows a minimum functionality. The functions or utilities can be grouped in four main categories. These are three-dimensional visualization utilities, cartographic design, cartographic modelling, and final display utilities. These should be accessible via a graphical user interface (Kraak, 1994).

Three-dimensional visualization utilities are concerned with the user's view of the map. They include geometric map transformations such as rotation, scaling, translation and zooming to position the map in three-dimensional space with respect to the map's purpose and the phenomena to be mapped. Geometric manipulations are necessary since in a three-dimensional image presented on a flat screen there is the possibility of elements disappearing behind other elements, as is shown in Figure 5.33. This can negatively influence the map's task of information transfer. To avoid this, a proper viewing position should be found by rotating the map around each of the x-, y- and z-axes separately. An important feature is the option to scale the map along the z-axis to find the proper vertical exaggeration. During the mapping process it should also be possible to use equipment such as stereoscope, to view the map in 'real' 3D.

Cartographic design utilities refer to the main design functions, which should include options to choose proper symbology. Amongst them are the definition of colours, line sizes, fonts, etc., and the positioning of legend, north arrow and scale bar. Information on the orientation of the non-orthogonal map in respect to the more familiar 2D view is also of importance. Referring to this orientation it should be noted that a direct link between symbols in the map and those in the legend is necessary. The design process is influenced by 3D perception rules. This means that as well as the use of graphical variables, pictorial depth cues, such as shading, texture, perspective and colour should be applied. The relative importance of each of the depth cues depends on the degree of realism of the final image. Other basic

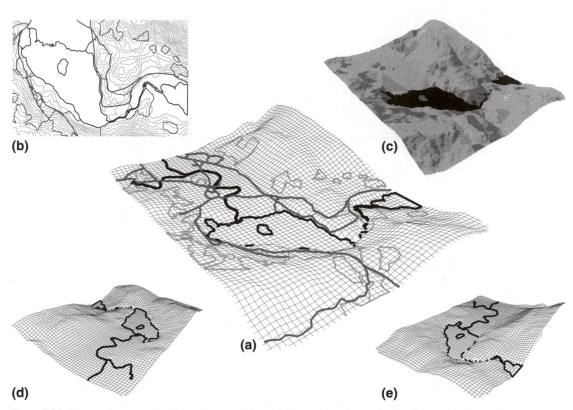

(b)

(c)

(a)

(d)

(e)

Figure 5.32 Perspective maps. Looking at the terrain from different viewpoints and with different data layers draped over

operations that might influence the design such as coordinate transformation, selection, classification and generalization of the data are assumed to be executed beforehand.

Cartographic modelling can be seen as manipulating maps or map layers. Tomlin (1990) describes it as a geographic data processing method, but within the framework of this book its only purpose is visualization. It also provides a link with GIS databases, and allows the cartographer or user to retrieve other map data and combine them partly or as a whole with the basic three-dimensional map data already displayed, as seen in the main map of Figure 5.33.

Final display utilities should help the user to produce the final map using known cartographic techniques as well as appropriate computer graphics techniques, both depending on the output medium (e.g. the monitor screen or Postscript). From the computer graphics world complete rendering programs can be implemented. These not only take care of hidden surface removal, texturing and shading, but also include complete atmospheric models

for realistic images. These last methods should only be applied if, from a cartographic point of view, they enhance the main task of the map, i.e. information transfer. At the moment GIS packages do not offer this type of functionality.

5.6   Topographic data: mapping and charting organizations

5.6.1   Introduction

Users working in a mature GIS environment will find most of the data needed in their own GIS database. If specific data needed for an analysis are not available the user has to consider other sources. When geometric data are needed, the local map library used to be the first place to go. After finding the map, it would be digitized. This might still be the only way of acquiring the data needed in many cases. However, it

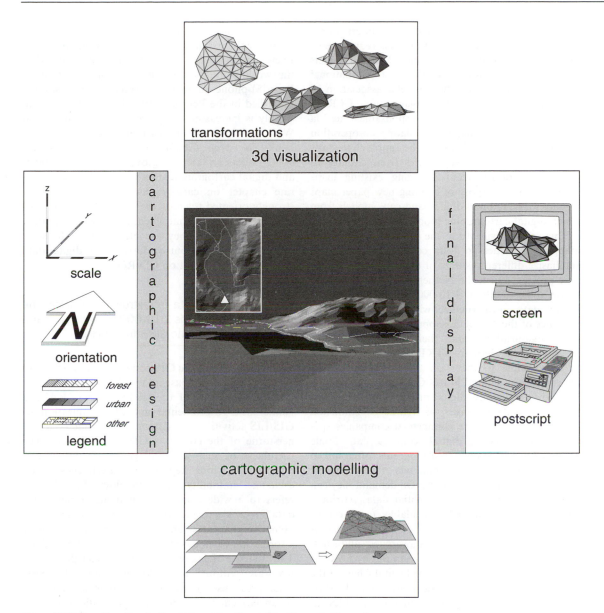

Figure 5.33 Functionality of a three-dimensional cartographic production unit

is much more convenient when the data can be obtained digitally. Government oganizations such as those responsible for topographic mapping and planning, have digital data for sale. For most topographic mapping organizations the sale of digital data is relatively new. For these organizations, the first task to be performed with computers was to implement the new technology in the conventional mapping process. Their goal was to produce the same maps using the new technology. The first data sets created were unstructured, and could only be used to draw maps. Such an approach is not that strange if one remembers their goal: the faster production of the paper maps. However, changing information needs in the civil and military market forced the mapping organizations to build structured topographic information systems. Organizations such as NATO and the European Union also increased the need for data sets that could cross international boundaries. Environmental problems, requirements of car navigation systems and digital terrain models are all examples of topics that do not stop at national boundaries. In Europe the umbrella association of the topographic mapping organization CERCO (Comité Européenne des Responsables de la Cartographie Officielle) is stimulating co-operation and standardization. However, the approach towards topographic mapping still varies from country to country. It ranges from digitizing existing maps with the single purpose of creating new paper maps (Figure 5.34), to build digital landscape models from which digital cartographic models can be derived.

This section will study the different topographic mapping activities in four countries, which each emphasize different aspects of the topographic mapping process. From the Netherlands, the role of the topographic database 1:10 000 within the national spatial data infrastructure will be discussed. The influence of the digital developments on British mapping organizations and their products will be described. The German ATKIS system will be used as an example to focus on the DLM and DCM approach. The United States Geological Survey will be used to show how digital products are marketed.

Regarding data sources, as well as governmental organizations there are commercial companies specializing in providing digital geometric data. Some of these have their own data programs, often linked to a GIS consultancy, with data sets of administative boundaries, road networks, etc. On demand they can provide any type of digital spatial data. Attribute data are often (digitally) available from census bureaus. Several GIS vendors have set up data programs as well, to increase GIS awareness and sales. An example is the ArcData program run by ESRI. It contains their own version of the Digital Chart of the World, and several other specific data sets. In their catalogue they also include geometric data files (in Arc/Info format), for sale by third parties, such as the Ordnance Survey or the USGS. Similar programs have started in several European countries. For instance, ESRI Germany recently released a CD-ROM called Arc-Deutschland'500, which is based on the German topographic map 1:500 000.

Many users might have trouble with the price of the data for sale. In which case, they should seek access to the electronic highway. Key words here are Bulletin Boards and Internet. These bring a wealth of public domain data within one's reach. It should be mentioned, however, that such data come as they are: no guarantees and no maintenance. But for many these data would just fit their requirements. Even some mapping and GIS software is offered in this world of public domain. A list of BBS (Bulletin BoardS) phone numbers and Internet addresses was published in the February 1994 issue of *GIS World*. Supply is increasing every day, especially of World Wide Web sites offering cartographic material. The book *The Map Catalog*, edited by J. Makower (1992), offers an interesting overview of many paper and digital cartographic data sources. It has a separate chapter on cartographic software, but it is strongly oriented towards the North American market. The problem is that when the data for the study area are available, they are often not at the scale needed, or they are out of date. Today digital data are mainly distributed on CD-ROM.

5.6.2 Spatial data infrastructure policies: the Netherlands 1:10 000 topographic database and the GIS infrastructure

In the Netherlands the GIS community is working to establish a core database at a scale of 1:10 000. This activity is the result of a feasibility study of RAVI, a Netherlands governmental umbrella organization for GIS/LIS activities. The database is considered a cornerstorne of the country's GIS infrastructure. The usefulness of such a digital geometric database is described by four key words: effectiveness, data exchange, suitability and technology. Effectiveness refers to a wide availability of digital topographic data for analysis and visualization purposes. Data exchange guarantees an easy exchange of geometric information when based on this consistent base data. The existence of a core database will deter those users who are tempted to build their own database. Widespread use of the same database will result in (financial) efficiency and savings in digitizing and updating activities. Currently, several country-wide databases exist, each with a different content, and each updated individually. The existence of the core database will improve the geo-information infrastructure, and will create many spin-offs.

Figure 5.35 shows the processes, materials and sources of the database. It is centred around the digital landscape model TOP10vector, created by the topographic survey of the Netherlands. It contains basic topography such as infrastructure, land use categories, and buildings, and will be updated every four years. As well as this topography, administrative

Plate 1   Basic Graphic Variables

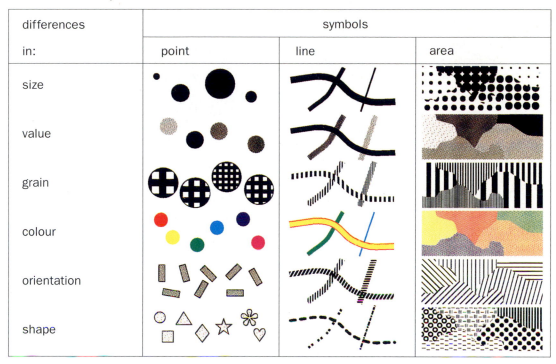

differences in:	symbols		
	point	line	area
size			
value			
grain			
colour			
orientation			
shape			

Plate 2   Differences in value or lightness

Plate 3   Differences in colour

Plate 4   Colour scale bar lengthened by adding saturation

Plate 5   Water hardness in the USA: gradation over and under a threshold value versus non-threshold choropleths

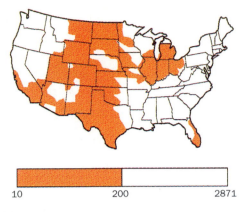

(a) binary

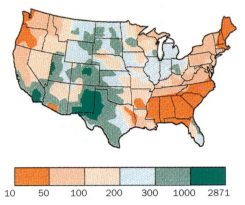

(b) diverging hues

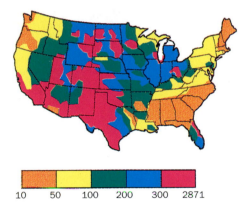

(c) spectral scheme colours

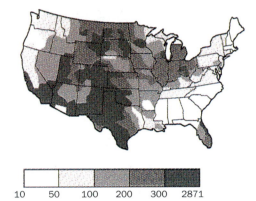

(d) sequential, no hue

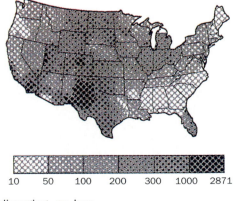

(e) diverging, no hue

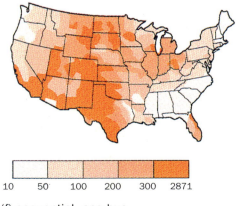

(f) sequential, one hue

Plate 6   Comparison of value and saturation

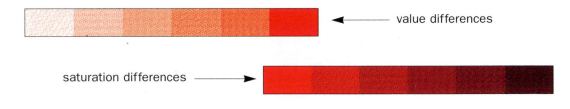

← value differences

saturation differences →

Plate 7   Trichromatic representation of relative values (after Bertin)

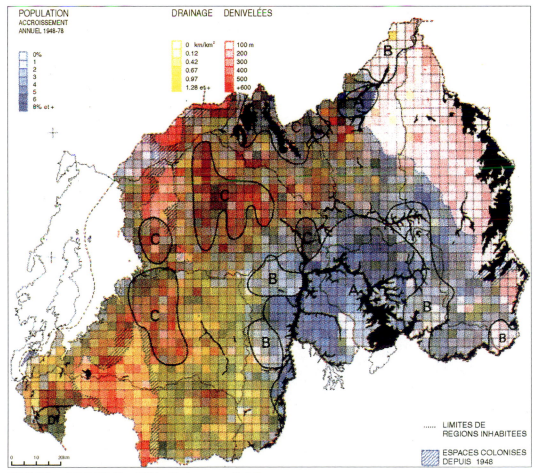

E.H.E.S.S. - O.R.S.T.O.M. - 1981 - Photogravure U.G.P.

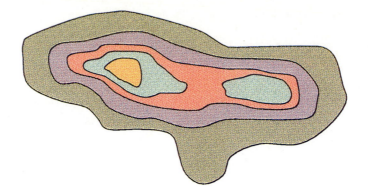

Plate 8   Layer zones with different colours for optimal retrieval of zones

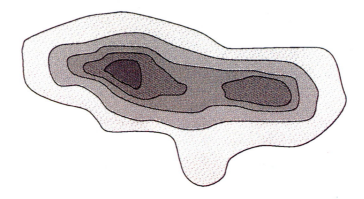

Plate 9   Layer zones with different tints for optimal recognition of trends

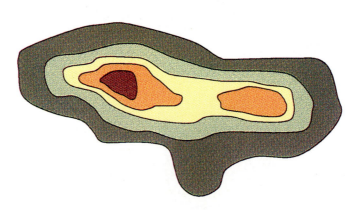

Plate 10   Layer zones with different convential colours

Plate 11   Primary colours

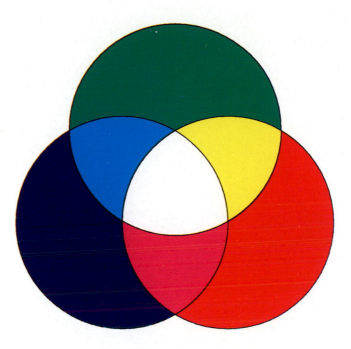

(a) additive colour scheme

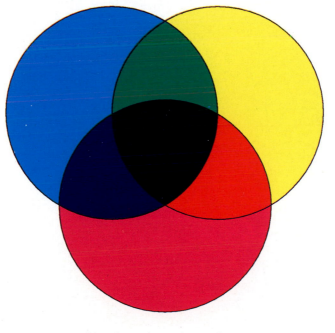

(b) subtractive colour scheme

Plate 12   Generalization related to atlas objectives

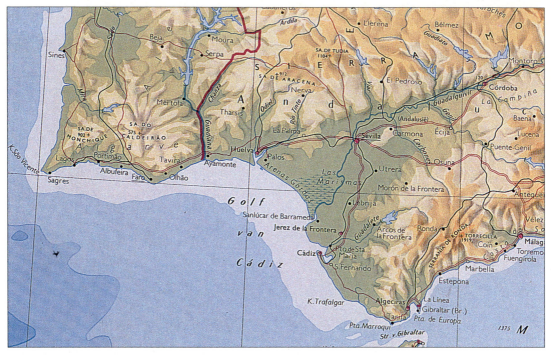

(a) From school atlas (Grote Bosatlas, 1995, courtesy Wolters-Noordhoff)

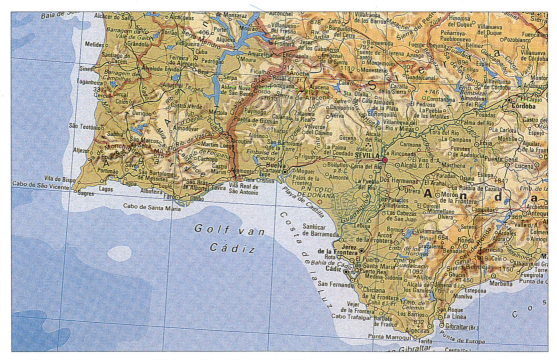

(b) From reference atlas (Wolters Wereldatlas, 1995, courtesy Wolters-Noordhoff)

Plate 13 Topographic map of the Netherlands 1:50.000, detail sheet 69W – 1991
(© Topografische Dienst Emmen)

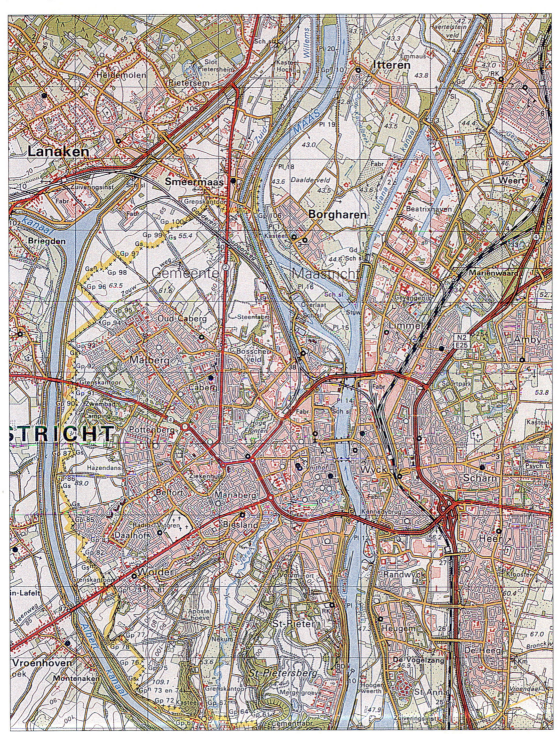

Plate 14   A page from an electronic atlas
(Mindscape's Multimedia Atlas version 5.0, 1994)

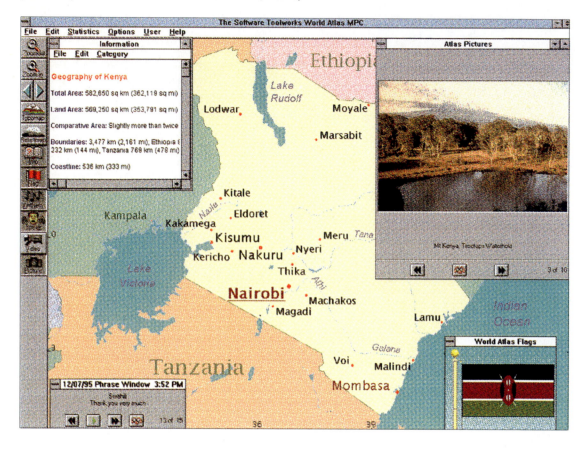

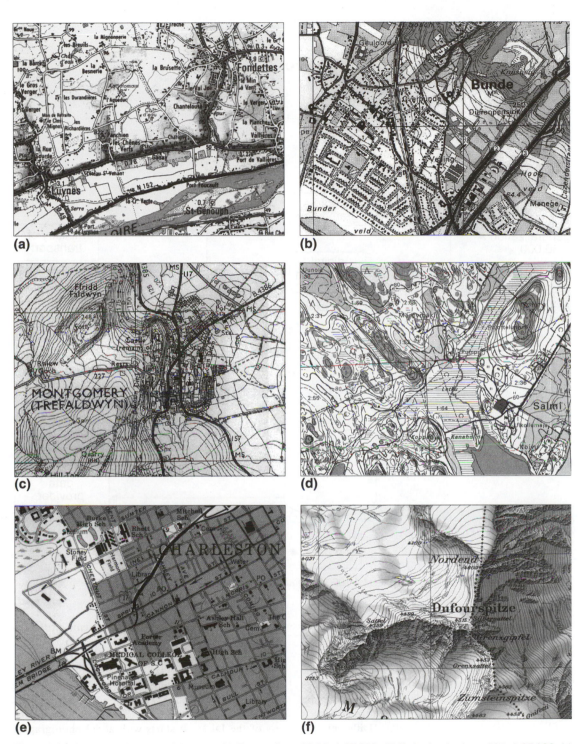

Figure 5.34 Examples of topographic maps: (a) France sheet 18-22, 1:25 000; (b) Netherlands sheet 69B, 1:25 000; (c) Britain sheet SO 29/39, 1:25 000; (d) Finland 2041-08, 1:20 000; (e) United States Charleston Quadrangle, 1:24 000; (f) Switzerland sheet 2515, 1:25 000

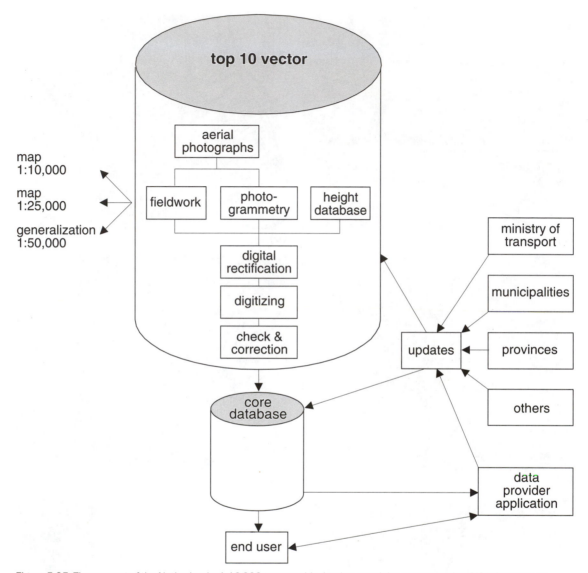

Figure 5.35 The concept of the Netherlands: 1:10 000 topographic database and the accompanying GIS infrastructure

boundaries and road centre lines are included. Both of these categories will be updated on a yearly basis. The data are available via several formal and ad hoc exchange standards, such as SUF2 (NEN1878 – the official Netherlands digital spatial data exchange standard), and Arc/Info and MGE GIS software packages. The GDF standard for road networks is available for the road centre line data. Different data suppliers, and even the users, work together to keep the database up-to-date. As well as the topographic survey, several other governmental organizations,

such as the state highway department, provinces and muncipalities, all contribute.

TOP10vector is a digital landscape model, originally designed to be used for digital cartographic models in the scale range 1:10 000 to 1:25 000. It will be completed in 1997. Figure 5.35 also shows (within the TOP10vector frame) the production process of the DLM. It starts with aerial photographs. Topographers go into the field with the photographs and classify the objects needed for the database. They also add information not visible in the photographs or on terrain, such as administrative boundaries and names. The photographs are scanned, and before the heads-on digitizing process can begin, the scanned photographs are rectified using a digital elevation model of the Netherlands. This model is based on an elevation map of the Netherlands at a scale of 1:10 000, and consists of a regular grid with a height value for every 25 m. After extensive checks and corrections the digitizing results in the DLM TOP10vector. The DLM is used to create the DCM for the 1:10 000 and 1:25 000 map series, and after generalization it is used to create the DCM for the 1:50 000 map series.

5.6.3 Landscape and cartographic models: the German ATKIS

Germany does not have a single topographic mapping organization responsible for all its maps. Up to a scale of 1:100 000 the individual states (Länder) are responsible. This includes the German 1:5000 large-scale map, and the map series at scales 1:25 000, 1:50 000 and 1:100 000. The federal mapping agency, IfAG, produces maps at scales 1:200 000, 1:500 000 and 1:1 million. During the 1980s the first steps towards ATKIS (Amtliches topographisches karto-graphisches Informationssystem) were taken. The goal of this topographic information system is to have digital topographic data available for all kinds of GIS applications. As with the Netherlands' core database, the principle is to collect the data only once, and have the database available for multiple use. ATKIS was the first topographic information system that introduced the terms 'digital landscape model' (DLM) and 'digital cartographic model' (DCM). Another goal of ATKIS is to effectuate a more rational and faster production of topographic maps by digital technology.

Currently, ATKIS is being produced at three scale levels, DLM25, DLM200 and DLM1000. The DLM25 is the responsibility of the Länder, and the others are the responsibility of the federal mapping agency. The basic principles of ATKIS are shown in Figure 5.36. From the figure it can be seen that the DLM is based on a topographic object catalogue. The objects and the relations between the objects model reality. Each topographic object is assigned its own object code. Apart from the code, its coordinates, object class, and attributes are stored. From the data model in the figure it can be seen that an object can consist of several object parts (point, line,

area), and can be part of a larger unit itself. Both raster and vector representations, as well as topological relationships, are stored. Depending on the visualization application, one or more DCMs can be generated. A symbol catalogue is available to store the proper graphical object representations in the DCMs

5.6.4 Small-scale customers: USGS and DMA's digital products

Spatial data supply is influenced by the policies of the different national mapping agencies. These policies may be widely diverging. In the United States, the United States Geological Survey (USGS) considers its spatial data as public property, which should be made available for a nominal charge to every citizen. However, a slight change towards a more commercial approach is evident. Great Britain and other European countries, on the other hand, regard spatial data as a commercial commodity, the access to which should be regulated by the market forces of supply and demand.

In the United States, civil topographic mapping is taken care of by the USGS. Map series at scales 1:24 000, 1:100 000, 1:250 000 and 1:1 million are produced. The country is in the process of establishing a National Digital Cartographic Database. Currently the database is filled with the digital version of the 1:100 000 map series. These data are also available as Digital Line Graph (DLG) and TIGER files. The last are the Bureau of the Census' adapted version of the 1:100 000 series, produced in order to collect and visualize 1990 census data. DLGs are digital cartographic models created by manually or automatically digitizing existing map sheets. The USGS classified their DLGs as large, middle and small scale. The large-scale DLGs are derived from the 1:24 000 maps. Here, it is interesting to note the different ways in which the terms 'large scale' and 'small scale' are perceived. Middle-scale DLGs are based on the 1:100 000 map series, while the small-scale DLG is based on the 1:2 million map series. The latter are distributed on CD-ROM. The DLG contains the following major map layers: boundaries (political and administrative), hydrography (streams, water bodies), hypsometry, transportation (roads and tracks, railroads) and cultural features. These DLGs are topologically structured and contain nodes, lines and areas. This approach guarantees spatial consistency (i.e. adjacent areas will remain

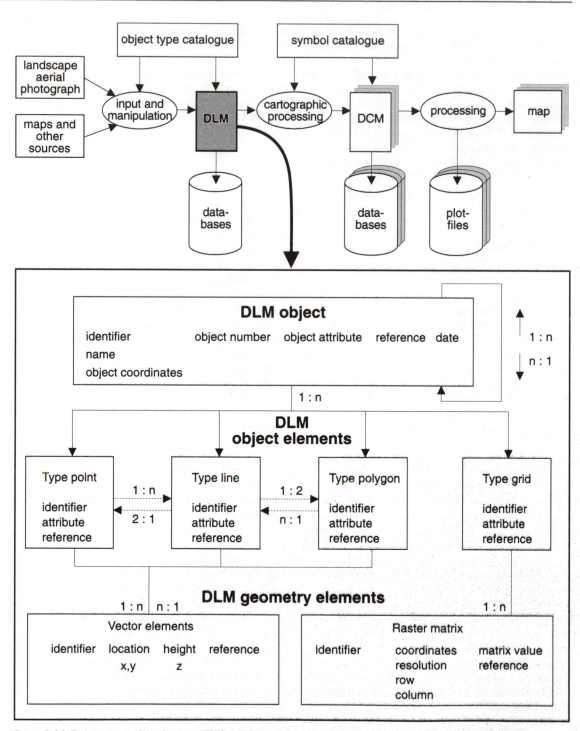

Figure 5.36 The structure of the German ATKIS database: the relation between DLMs and DCMs (Grünreich, 1990)

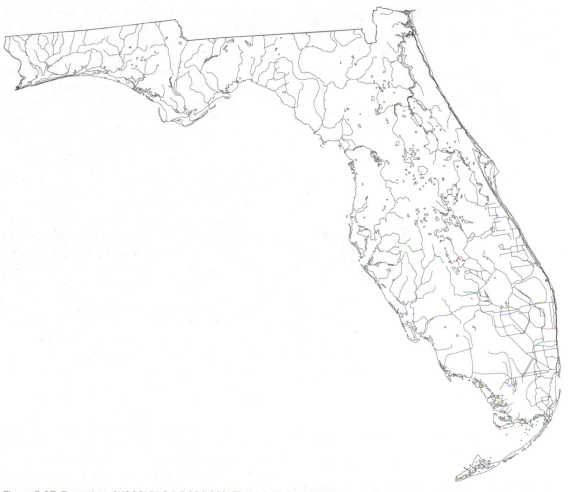

Figure 5.37 Examples of USGS' DLG 1:2 000 000: Florida's drainage layers

adjacent and connecting lines will remain attached to each other). Attribute information is available as well. In the US the DLG format function as a *de facto* exchange standard. Figure 5.37 shows an example for Florida.

Digital Elevation Models (DEMs) are also produced by the USGS. They contain height values stored according to a regular grid, and are produced on the basis of the 1:24 000 and 1:250 000 map series. The DEMs derived from the 1:24 000 are called 7.5 minute DEMs, since the map sheets covered by the DEM files cover 7.5' by 7.5' in the geographical coordinate system. After photogrammetric techniques the interpolation of the contour lines and height points found on the maps is the most important method used to build these DEMs. The coordinates, in UTM format, are stored in profiles, with points every 30 m. The accuracy of the DEMs is strongly dependent on the data source and interpolation technique applied. The vertical accuracy of the 7.5 minute DEM is at least 15 m. Based on the 1:250 000 map series, a 1° DEM is available.

In 1992, the Defence Mapping Agency (DMA), in close co-operation with its Canadian, Australian and British counterparts, issued the *Digital Chart of the World* (DCW) 1:1 million (see also Section 2.4). Figure 5.38 shows the contents of the hydrography layer for Cumbria.

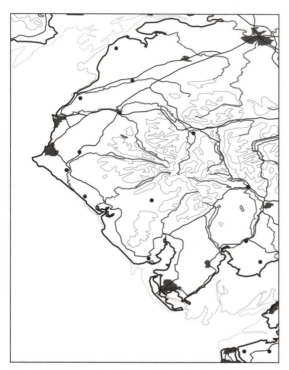

Figure 5.38 Examples of several layers for Cumbria of DMA's DCW; layers of populated places, drainage, hypsometry, roads, coast/political boundaries

5.6.5 Large-scale customers: British National Grid Plans

The Ordnance Survey is the topographic mapping organization in Britain (Figure 5.39). It is responsible for the small-, medium- and large-scale mapping of the country. Small- and medium-scale map series are produced at scales 1:625 000, 1:250 000, 1:50 000 and 1:25 000. More than 220 000 large-scale maps, called National Grid Plans, are published at scales 1:1250, 1:2500 and 1:10 000. To speed up and enhance the production of these large-scale maps, computer cartography was introduced at the beginning of the 1970s. After digitizing more than 40 000 maps, times changed. During the mid 1980s two key words had a great impact on the approach: GIS and privatization. The first affected the cartographic approach: one had to change from unstructured drawing files to topological structured data. The second had economic consequences: income and savings. The combined impact completely changed the procedures. The Ordnance Survey started to build its

Topographic Information System (Figure 5.40). Its purpose was to provide an infrastructure to collect, store, manipulate and supply structured topographic data. During this build-up several conditions were considered. The spread of GIS would require an increase in structured digital topographic base data. If the Ordnance Survey intended to sell these data they should meet the needs of the customer. However, an increase in digital sales would probably decrease the demand for paper maps. At that time it was decided that 1995 would be a critical year since potential customers had indicated that this would be a deadline: if the data were not available by then, these customers would start their own mapping activities. The analogue–digital operation should be efficient, and should incorporate facilities to keep the digitized data up to date. A field update system, executed locally, was introduced. It guarantees that digital data will never be more than a fortnight out of date.

Today the large-scale programme has been realized for most of Britain. All of the urban areas are available. The user can obtain the data digitally, but can still get paper maps on demand. These paper maps produced on demand are known as Superplan. The basic idea of Superplan is that the consumer can buy a customized map around the corner. In London, for instance, one can visit Stanford's map shop and have a map sheet plotted either centred on one's own house, or for a certain postcode area. The contents can be specified as well, since the database is structured in layers. Output is currently limited to black and white plots or films, in sizes between A4 and A0. Figure 5.41 shows some Superplan characteristics. In the shop the direct environment is stored locally, and updated on a daily base. If needed, the customer can also get maps of other parts of Britain, via a direct link with the main database in Southampton, where the Ordnance survey's headquarters are situated.

Section 5.6 has described several procedures for processing and distributing spatial data. It will be clear that there are differences in data structures, projections, datums, even ellipsoids. These all prevent proper integration of data. That is why, on a continental level, a European Triangulation network has been set up, which allows one to convert coordinates from one national system into another, by allowing a transformation into ED (European Datum) coordinates. In turn, the coordinates from the ED system can be expressed in the WGS 84 system (WGS stands for World Geodetic System).

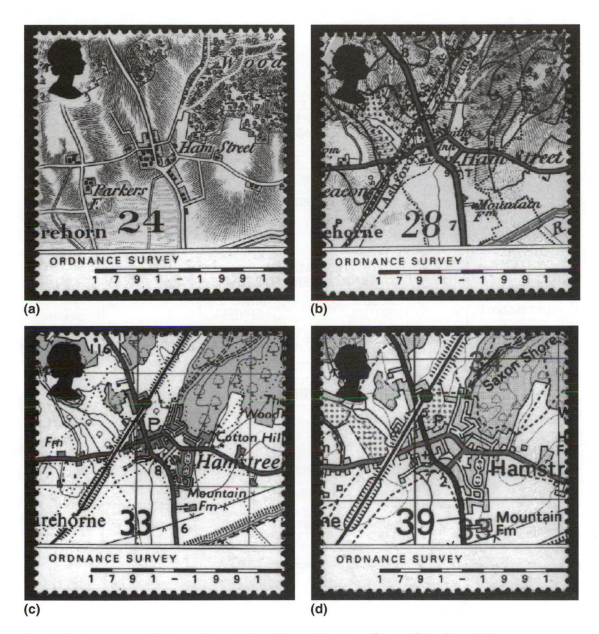

Figure 5.39 Development of Ordnance Survey products illustrated by stamps (Courtesy British Royal Mail)

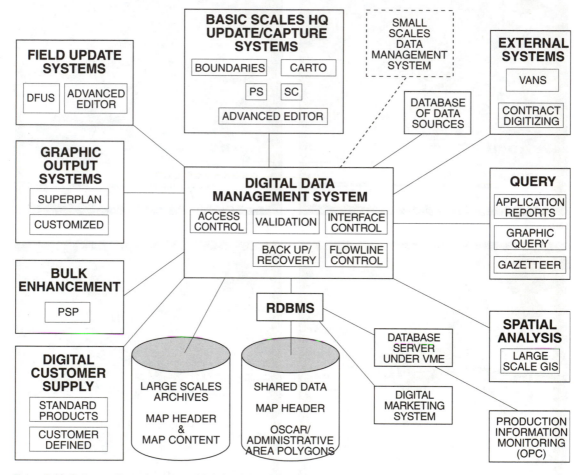

Figure 5.40 Ordnance Survey's topographic Information system (from Rhind, 1992)

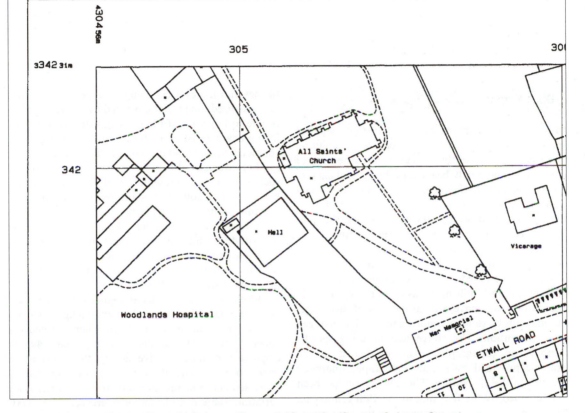

Figure 5.41 Large-scale mapping at the Ordnance Survey and Superplan (Courtesy Ordnance Survey)

Further reading

Brassel, K.E. and R. Weibel (1988) A review and conceptual framework of automated map generalization. *International Journal of GIS*, **2**(4), 229–244.

Buttenfield, B.P. and R.B. McMaster (eds) (1991) *Map generalization. Making decisions for knowledge representation.* London: Longman.

Canters, F. and H. Decleir (1989) *The world in perspective: a directory of world map projections.* Chichester: John Wiley & Sons.

Imhof, E. (1982) *Cartographic relief representation* (English translation). New York: W. de Gruyter. First published in German in 1965.

Maling, D.H. (1992) *Coordinate systems and map projections.* Oxford: Pergamon Press.

Raper, J. (ed.) (1989) *Three-dimensional applications in GIS.* London: Taylor and Francis.

Snyder, J.P. (1987) *Map projections – a working manual* US Geological Survey Professional Paper 135.

Snyder, J.P. (1989) *An album of map projections.* US Geological Survey Professional Paper 1453.

Snyder, J.P. and H. Stewart (1988) *Bibliography of map projections* US Geological Survey Professional Paper 1856.

CHAPTER 6   Map design

6.1 Introduction

In the previous chapters we have seen that maps are
spatial images that can influence people's conception
of space. Maps have this influence partly because of
convention and partly because of the general characteristics of the graphic cues used. Convention especially plays a role in topographic mapping: most of
the symbols used on topographic maps (see Chapter
5) have come down to us in a form conditioned by
18th century examples and we have stuck to them
ever since. Amongst these conventions are that
water is represented by a blue colour, forests by a
dark green, and built-up areas by red, grey or pink.
Association may have been at the root of this usage,
but may not be valid any more, and so it has changed
into convention. The convention of using specific
symbols on topographic maps originated in the
example provided by French topographic mapping
practice in the 18 century. This convention has been
strengthened by the fact that in the 19th century all
topographic maps were produced for the same objective, i.e. infantry warfare.

The result, for topographic maps, is a large collection of symbols – for buildings, infrastructure, terrain aspects, hydrography and administration – that
has been more or less standardized, and that works
because it has been standardized. It works moreover
because people are used to this kind of symbology on
these kinds of maps; it can be learned by those who
use topographic maps.

There is an ever increasing proportion of maps,
however, that have nothing to do with descriptions
of the terrain and its fixed assets, and instead have
other objectives, i.e. the thematic maps defined in
Section 3.2. Here, because of the ever changing
themes and the ever changing aspects of reality that
are visualized, one is not governed by convention but
is able to improve information transfer by using the
innate characteristics of the variation in the graphic
characteristics (e.g. shape, colour, size, texture) of the
symbols we use. When we study map symbols as
such, i.e. not as a representation of the earth's surface
but as a set of dots, dashes and patches, we find that
it is this variation in graphical aspects which conveys
meaning to the map reader, like a sense of varying
magnitude, or differences in nature. It was the French
geo-cartographer Bertin who placed the various sen-
sations created by the variations in graphical aspects
in a logical strucutre (Bertin, 1983).

Indeed, when one looks at Figure 6.1(a), one will
perceive a dark circle against a light background, but
this will not tell us much. Even when a legend is
added to indicate that the circle represents 40 000
employees in an automative factory, this information
seems to be a waste of space, as it could have been
expressed better in alphanumeric form. As there is no
variation, there is no frame of reference, no context.
But as soon as this single symbol is put into a geographic and data context, as in Figure 6.1(b), this
variation in graphical cues will immediately render
spatial information more effectively than an alphanumeric description could. Differences in size are
obvious and will be perceived automatically as differences in number (of employees in automotive factories). So a hierarchy will be perceived based on
these differences in number. As well as the hierarchy,
a pattern will be discerned, influenced by the relative
distances between the various symbols.

So the differences in symbol size are an important
characteristic, conveying to the map reader the sensation of differences in number – in our example here,
to convey some idea of the relative importance and

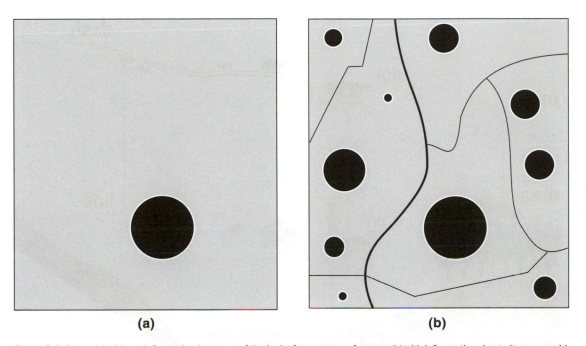

(a)   (b)

Figure 6.1 A map (a) without information because of the lack of context or reference; (b) with information due to its geographical and thematic context

distribution of the automotive industry in an area. It is on these automatic reactions to variations in graphical cues that graphical grammar is based (see Section 6.3).

So, as the symbology of topographic (or 'inventory') type maps covered in Chapter 5 is based more on convention than on this cartographical grammar, and as thematic, communication-oriented maps (such as those discussed in Chapter 7) are more based on this grammar, it is here in Chapter 6 that this grammar will be discussed. After descriptions of data gathering techniques (Chapter 2), map functions (Section 3.4) and base maps (Chapter 5), it is here that the characteristics of the graphical signs will be explained. These perceptual characteristics of the graphical signs will have to be matched to the data characteristics analysed in Section 7.2 and the communication objectives (Section 7.4) in order to satisfactorily portray the information requested.

6.2 Symbols to portray data related to points, lines, areas and volumes

The data that have to be visualized will always refer to objects or phenomena in reality. These can be heights measured at specific points, traffic intensities measured along a route network, numbers of inhabitatnts living in an area, or the volume of a hill in thousands of cubic metres. In Section 7.2 we refer to specific aspects of the data; here we will show the graphical means we have at our disposal, to represent them.

In cartography we use dots, dashes and patches to represent the location and attribute data of point, line, area and volume objects as in Figure 6.2. (One could mention here that the definition of point, line and area objects, i.e.: objects that refer to point, line and area locations, is a matter of scale: a line which represents a river would have to be exchanged for an area if the scale of the map was to increase. The built-up area of a settlement would be rendered by a dot if the scale of the representation were to decrease enough.) It seems to be obvious that point data are represented by dots and that area data are represented by patches, but there is more to it than that.

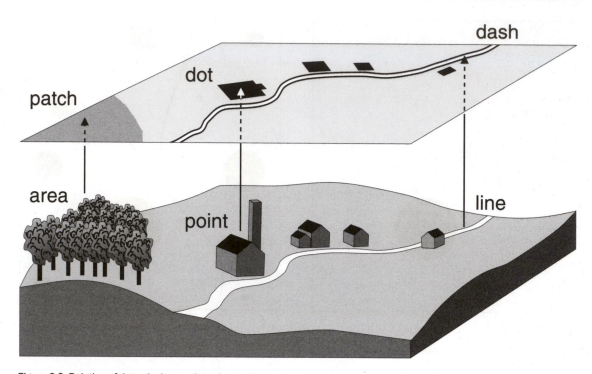

Figure 6.2 Relation of dots, dashes and patches to the point, line, area and volume objects which they can represent

Figure 6.3 provides an example of various kinds of point data: equally sized dots that each denote the same value (e.g. 10 inhabitants), dots that vary in size and thus represent different quantities for specific point locations (as in Figure 6.2) but – and this is represented by the addition of boundaries – these proportionally sized dots can also refer to enumeration areas. In the latter case the dots could be considered area symbols, even though each of them is centred on a point location (e.g. the respective area's gravity point). Another application of dots for rendering areal data is in a regular grid mode. The value valid for an area can be assigned to the nodes in a regular grid superimposed over the area (see Section 7.5.5 and Figure 7.30).

Figure 6.4 presents some dashes used to express various types of linear data: boundaries, roads and railways, flowlines proportional to the number of passengers transported, etc. Lines can also be used to represent areal data, by using them as shading, but, as is the case for point symbols, they must be combined in such a way that they are perceived as patterns and not as individual points or lines. Lines can also be used to indicate volumes. In the same

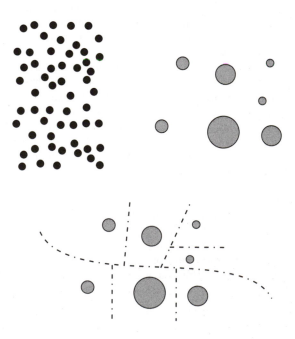

Figure 6.3 Various types of dots that are used as point or area symbols

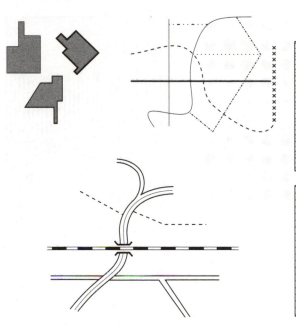

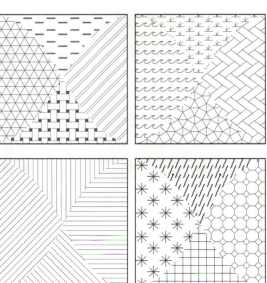

Figure 6.4 Various types of dashes that are used to symbolize linear objects

Figure 6.5 Patches or patterns used to symbolize area objects

way, a string of dots representing a linear feature could be referred to as a line symbol.

Figure 6.5 illustrates a number of patches used for representing areal data: patches that suggest qualitative or quantitative differences between the various areas concerned. As said above, it is their repetition that leads us to perceive the dots and dashes as area symbols. In all the cases in Figure 6.5, within the boundaries of each area the patterns are homogeneous. If patterns were not homogeneous, they could be used to indicate volumes: hill shading would be a good example (Figure 5.30).

6.3 Graphic variables

As in the example in Figure 6.1, it is the difference in size which map readers perceive as a difference in numbers. In order to systematize the perceptual characteristics available, we will list them here:

- difference in size
- difference in distance
- difference in order
- difference in quality

When confronted with differences in (grey) value or lightness of tones, one will experience a sensation of perceiving differences in distance: in Figure 6.6(a) the population density is represented by grey tones that have an increase which is perceived as regular: the distances between the classes are similar. That is why, for most percipients, these population density values would show a regular increase.

This same difference in population density value (i.e. the difference between successive values) can also be perceived from point symbols different in size that are applied as area symbols in grid-type maps. These would have the added characteristic of also allowing ratios to be perceived (this density is so and so many times higher than that density; Figure 6.6b).

Differences in order will be perceived from differences in symbol size, from differences in grey value or lightness and from differences in texture (Figure 6.6c) as well as from differences in colour saturation. Nominal or qualitative differences will be perceived from differences in colour hue, shape and orientation.

In order for the symbol differences to be perceived as qualitative differences only, they must be perceived as having similar values. If one colour were to be perceived as much darker than another, then order

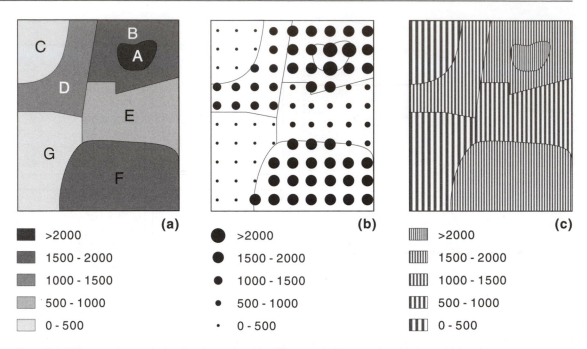

Figure 6.6 Differences in population density rendered by differences in (a) grey value, (b) size and (c) grain

differences would be experienced as well, the darker colour denoting areas that would be both different and more important than the lighter areas. In practice, darker colours can only be used to represent qualitative information for small areas; otherwise they dominate the image too much. In Figure 6.5 the two lower examples aim at showing different qualities only, as do the patterns in Figure 7.23.

While discussing these perceptual characteristics of the graphic cues, we encounter various basic differences in the graphic character of the symbols we discern. All the differences imaginable between symbols can be summarized as being cases of six graphical variables. Bertin (1983) discerns, as basic graphic variables (Plate 1),

- differences in size
- differences in lightness or (grey) value
- differences in grain or texture
- differences in colour hue
- differences in orientation
- differences in shape

Figure 6.1 is one example of differences in symbol size but differences in line width or in areal symbols like proportional dots in grid patterns (Figure 6.6b) also qualify (the graphic variable 'differences in size'

never refers to the surface of the areas the symbols refer to!).

Differences in (grey) value or lightness are exemplified in Figure 6.6(a) and also in colour Plate 2.

With differences in grain or texture, Bertin referred to differences that emerge when a specific pattern is being enlarged or reduced. The ratio between the areas that are black and white respectively will remain the same during this photographic process, but at the same time the coarser the pattern, the higher it will be perceived in the resulting hierarchy. Figure 6.6(c) shows a map using differences in texture in order to generate an impression of order amongst the categories discerned.

Differences in colour hue only work in providing qualitative differences when they are perceived as having similar lightness. Totally saturated colours (i.e. when the whole area is only covered with ink with one specific wavelength, so is not mixed with either white, black or any other colour hue) have different lightness values. Plate 3 will show this.

Differences in orientation refers to patterns and not to the line elements that form the base map. The lower left example in Figure 6.5 shows differences in orientation. These can either refer to line patterns or to dot patterns.

Differences in shape can refer to differences in the dots, in the lines, or in the patterns used for area symbols. Again, shape differences as a graphic variable would never refer to the shapes of the areas the various colours, patterns or symbols refer to – they only include the symbols themselves.

As well as these six graphic variables discerned by Bertin (1983), there are some extra ones proposed by North American cartographers. These would be differences in colour saturation, in arrangement and in focus. Differences in arrangement refers to the regularity or non-regularity of the distribution of symbols. Focus refers to the clarity with which the symbols are visible, and so to their definition on the plane. We will not use arrangement or focus here because they can also be thought of as referring to the base map. Colour saturation (also called chroma) can be defined as the percentage of the reflection of light from an object composed of colour of a specific wavelength. The larger the reflection percentage of the light with this wavelength, the more saturated or brilliant this specific colour will appear.

The importance of discerning these basic graphical variables and the perceptual characteristics of the differences in each of them is that they help map designers in selecting those variables that provide a sensation which matches the characteristics of the data or the communication objectives. Figure 6.7 relates graphic variables and perception characteristics to each other, and it is the key illustration of this chapter. As can be seen, it also refers to the dimensions of the sheet of paper or monitor screen on which our maps are drawn.

These dimensions of the plane also have perceptual characteristics: as one location is not equal to another, space differentiates. Because of contiguity, individual point objects can be grouped. If on the way from A to C one has to pass B, there is a distinct order, A–B–C, which cannot be changed. Distances or angles measured on the ground have numerical connotations, and can have both interval and ratio aspects.

It would perhaps be expected that variations in size, colour value and texture would also be able to denote nominal differences (Figure 6.7). They can, but at the same time they have hierarchical connotations that dominate overall impressions. That is why these fields have been left blank in the matrix.

It is important that not only can the correct impression be gained, but that it is gained with a minimum of exertion. Here we may introduce the concept of visual isolation (which Bertin calls sélection), which indicates whether or not all the relationships that can be perceived between the various categories discerned on the map can be perceived at a glance. Not all graphical variables work equally well in this respect, depending partly on the number of categories one wants to be differentiated on the map.

If it is nominal differences that one is interested in, with areal objects (patches) different in colour, eight classes is the maximum that can be differentiated; if more than eight classes are selected, it will no longer be possible to discern the distribution of each of them anymore. So, if five classes of linear objects are to be distinguished between at a glance, then size, colour hue, value and texture are to be selected, according to Figure 6.8. If the data to be represented are ordered, colour hue would not qualify any more. If the pattern were dense, there would be no place for lateral extensions, so size could be discounted. Whether texture or (grey) value were used would be left open to one's personal choice.

6.3.1 Visual hierarchy

The selection of the most suitable graphic variables and processing them according to the proper mapping method is still not enough from a map design point of view. Legibility considerations play a role,

	qualitative	ordered	distance	proportional
dimensions of the plane	X	X	X	X
size		X	X	X
(grey) value		X	X	
grain/texture		X	X	
colour hue	X			
orientation	X			
shape	X			

Figure 6.7 Relation of graphical variables to perceptual characteristics (based on Bertin's *Semiology of Graphics*, 1983)

	dots	dashes	patches
size	4	4	5
(grey) value	3	4	5
grain/texture	2	4	5
colour hue	7	7	8
orientation	4	2	-
shape	-	-	-

Figure 6.8 Visual isolation: the number of categories that can be perceived at a glance

and these can be subdivided into contrast, graphical density and angular differentiation.

Introducing contrast is based on the assumption that the map data will consist of a number of categories that will each have a different role to play in the spatial message. The data analysis process (Section 7.2) will result in the identification of more and less important data categories. The examples in Figure 6.9 both show the number of employees in the service industries in the Netherlands. Figure 6.9(a) does so poorly. Although the map does portray the numbers of employees per province, it is the sea which is most conspicuous, followed by the surrounding countries. Both the Netherlands and the proportional circles score equally low in their conspicuity. The map in Figure 6.9(b) shows how the information hierarchy (the sequence of most to least important aspects of the data to be shown) should be portrayed. It is the employees that should stand out most, followed by the province they work in; the surrounding areas should come last. One need not even differentiate between the sea and the neighbouring stages, as they are equally unimportant to the map theme (unless we expect the proximity of either of them to influence the distribution of the employees in this sector).

6.3.2 The use of colour

There is more to colour in maps than its suitability for distinguishing nominal categories. Amongst the aspects of colour that are differentiated between are colour hue (the dominant wavelength), colour saturation (the proportion of the light reflected which con-

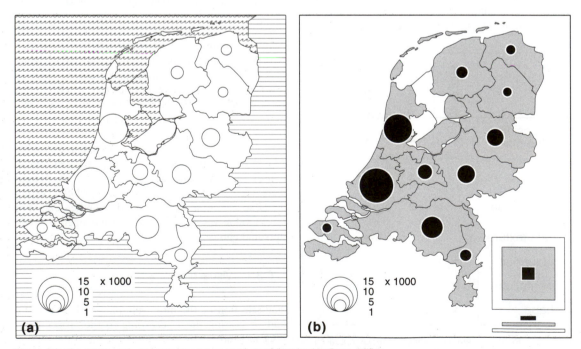

Figure 6.9 Graphical or visual hierarchy: (a) poor; (b) good (inspired by Dent, 1985)

sists of this particular wavelength; this can be diminished by adding white or black or other colours), and the (grey) value or lightness (the impression it would give when shown on a black and white monitor). The number of different (grey) values that can be discerned within one colour depends on its hue: for yellow only three steps can be discerned, while for red and blue six or seven can be distinguished.

Colour perception has psychological aspects, physiological aspects and connotative/subjective and conventional aspects. Amongst the physiological aspects it has been noted that on small areas it is difficult to perceive colours, and that between some colours more contrast is being perceived than between others (so this combination could be used in order to improve acuity).

Saturation differences can be emphasized in practice by adding black screens to the colour. This will often look like a pollution (see Plate 6), but its effect is that a colour scale (a number of classes differentiated on the basis of variation in colour value) can be lengthened. The figure in Plate 4 shows a colour scale that was lengthened by adding saturation differences to it. Colour differences are also used in situations where deviations from a central situation are indicated. The simplest case would be a binary map, showing those areas/points below or above a central value or threshold value. Several ways of showing gradations above and below this threshold value are possible (Plate 5; see also Brewer, 1994).

Bertin has done the same with three colours that represent data which together make up 100% of the observations (Plate 7). As can be seen from the resulting maps, it is easiest to find the area one lives in in the binary map (Figure 6.10); general trends are best seen from the non-diverging scales (Figure 6.11, mid-

dle), and the coloured maps seem to perform better than the non-coloured ones.

If it is individual values we would want to show a subdivision into different colours would be best. This is in fact a default setting in some GIS programs, but, as can be seen from the maps in Plate 8, no spatial trend is visible from the map thus coloured in, contrary to a map with value differences (Plate 9) or, to some degree, conventional layer zone colours (Plate 10).

How do we get the same colours that are selected on screen onto the final printed or plotted map (Brown and Schokker, 1989)? The problem here is that the colours on the monitor are additive colours – like those of all other light sources (sun, neon lights, etc.). The colour hues we see are perceptions of the particular wavelength radiation from these sources. In Plate 11 one can see that the primary colours emitted from these sources are red, green and blue. When there is an overlap between these primary colours' wavelengths, the addition of them leads to the perception of secondary additive colours: yellow, magenta and cyan. And when all three primary colours are added, this will lead to a sensation of white. The more wavelengths added together, the brighter the image will be.

In contrast, printing or plotting inks, when applied to paper, act like filters. Therefore, they are called subtractive colours. A paper sheet printed cyan will absorb the red wavelengths and only reflect blue and green, which together (Plate 11, subtractive colour scheme) will give us the impression of cyan. One can see that it might be problematical to convert the additive colours from the monitor into subtractive colours on a printed or plotted map.

The colours on the screen are the result of red, blue or green phosphor dots being activated by an elec-

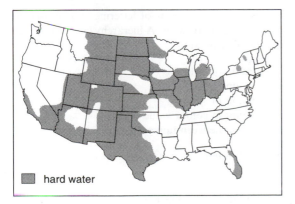

Figure 6.10 Binary map: water hardness in the United States

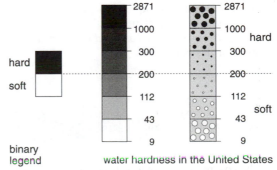

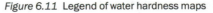

Figure 6.11 Legend of water hardness maps

tron beam. The more electrons emitted, the more intense the light of the phosphor dots will be. When activated by the respective electron beam, a green dot will emit a green radiation; when not activated, it will not radiate any colour. When one perceives colours on the monitor screen, it is a result of the differences in intensity of light emitted by green, red or blue phosphor dots.

Printed or plotted onto paper, variations in colour will be the result of varying amounts of dye or of the size of the dots printed in cyan, magenta, yellow or black. The dots can be superimposed or they can be printed next to each other, so that dots in various colours are intermixed. This is known as dithering, and will result in the eye perceiving shapes in an average colour, which is less acute than when the dots are overprinted.

In order to get a really good fit between screen and plot colours it is wise to produce colour charts (Brown, 1982). These are sheets with little square of different colours, resulting from different intensities of yellow, magenta and cyan, with which the screen colours (which are red, blue or green-based) can be matched. The purpose of these colour charts is to judge in advance what the colours on the screen will look like when printed.

Some colours (red, yellow) seem to come forward from the paper they have been printed on, others (green and blue) seem to retreat. This rather weak phenomenon was once considered so important that the layer zones in height representation were coloured according to this phenomenon: bluey-green for the area below sea level, green and yellow for lowland and hills and brown colours for mountains, with red for areas over 5000 m. The effect was meant to be that those areas that appeared to be nearest to the viewer were in fact those farthest away from sea level. This effect, however, was impossible to measure then – but by the time this was realized it had already been applied in school atlases and had become so popular that it had become a convention. The *Times Atlas of the World* provides a prime example of this convention.

6.4 Text on the map

The technical procedures of applying text to the map will be covered in Chapter 8. It is here that conceptual and design aspects will be dealt with first. By the phrase 'text on the map' we mean the text within the map's frame, and not the additional information (title, legend, etc.) in the map's margin. Text on the map itself has the primary function of providing spatial addresses – by naming the various map objects (geographical names or toponyms are used for that purpose; Kadmon, 1992). A secondary function would be to indicate the nature of objects. On topographic maps, terms like 'factory', 'cemetery', 'airfield', etc., are used for this purpose.

When compared to text in books, text on maps has some special characteristics. Map text consists of individual words instead of sentences; the words are unfamiliar instead of familiar; and there might be larger spacings in between the letters than is customary in book text. In contrast to book text, names on maps do not have to be horizontal, and they certainly are not neatly placed in lines: there is a jumble of different styles and sizes; the words refer to symbols instead of to each other as is the case in book text and – and this is the worst aspect – the text on maps superimposes lines and patterns. For all these reasons text on maps has some extra requirements.

They should be easily identifiable and legible, even if larger interspaces apply. It should be possible for the lettering styles selected to be differentiated between through differences in boldness and size. If these requirements can be met, the next requirements for the selection of letter types are as follows:

- they should be able to convey hierarchies (differentiating between more and less important objects or object categories);
- they should be able to show nominal differences (between different data categories);
- it should be possible to use them for relating to point, line and area objects.

Let us look how aspects of lettering can be used to meet these requirements. A hierarchy can be achieved in a number of ways:

- variation of boldness
- variation in size
- variation in spacing
- variation in colour value
- variation between upper and lower case

Nominal differences can be created by

- variation in colour
- variation in style (shape)
- variation between roman script and italics

Figure 6.12 Variation of map scripts in order to show hierarchical and/or nominal differences

Figure 6.12 illustrates these differences and Figure 6.13 is an application of the differences. Frequently, in order to achieve a hierarchy, a combination of the above-mentioned variations is used, resulting in, for instance, a series that goes from lower case italics, 3 mm high, to bold roman upper case letters, 10 mm high, to differentiate between settlements on the basis of number of inhabitants.

These nominal and hierarchical differences are important because they help one to find specific names on the map. In the legend one will always be able to find out how specific object category names are shown on the map. If one finds that a name belonging to a category (e.g. a water name or hydronym, or a place name or the name of a physical object such as a mountain) is displayed using a specific script type, one will know what shape of name to search for on the map, so the time needed to find this name on the map will be substantially reduced. This is because only names given in this specific script type will be targeted visually.

For printed maps a requirement for map text would be that it is visible without having to use a magnifying glass; should not be too thick (and risk obstructing map detail) or too thin (and risk the danger of being lost in the map detail). There should be good differentiation between the letters e and c, between a and u, and between u and v, between 3 and 5 and 8, and between 1 and 7. Lettering should be resistant against zooming in/out to a reasonable degree.

Different objects on the map have different requirements: point objects such as cities are preferably named by text that is placed slightly above or below the line the object symbol is on, and preferably to its right. Linear objects such as rivers are, on paper maps, preferably named by text that is parallel to and close to the lines, and even follows their bends (this is something difficult to achieve with current mapping software). In order to relate to area objects, one would try to show the extent of the object by covering its largest extension with its name, which means both interspacing and, when this largest extension is not horizontal, tilted names.

It is with these techniques that an optimal relation between text and map will emerge.

Finally, in order to promote rapid identification, a short note about title and 'vedette'. From the data analysis procedure covered in Section 7.2, the invariant aspect (the information aspect common to all data elements) will be deduced. This common aspect can be expressed as the title of the map. It is customary to represent the most important aspects of the title (those refering to area and theme) as prominent key words, called 'vedette' in French (e.g. 'Land cover in South Sumatra'), and this will be displayed boldly on the map. The complete title will be rendered less conspicuously underneath (e.g. 'Land cover as interpreted from SPOT imagery taken in the period 1990–92 of Sumatra south of the 4th meridian'). This will help in speedy identification.

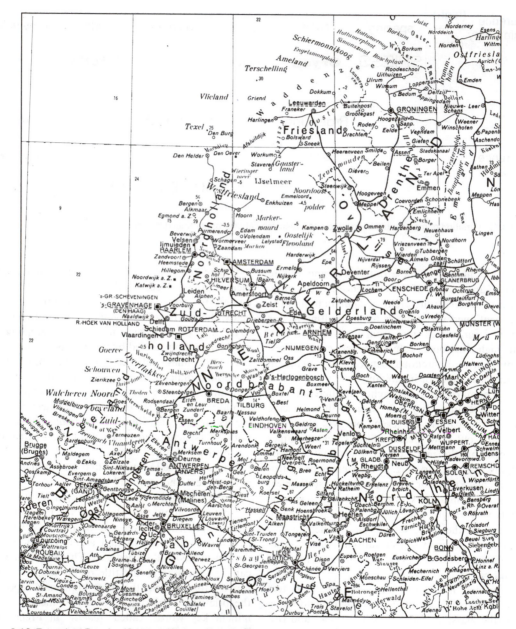

Figure 6.13 Type plate Benelux (Courtesy Wolters-Noordhoff)

Further reading

Bertin, J. (1983) *Semiology of graphics*. Madison, WI: University of Wisconsin Press.

Brewer, C.A. (1994) Color use guidelines for mapping and visualization. In A.M. MacEachran and D.R.F. Taylor (eds), *Visualization in modern cartography*. Oxford/New York: Pergamon, pp. 123–147.

Brown, A. (1982) A new ITC colour chart based on the Ostwald colour system. *ITC Journal*, **1982**(2), 109–118.

Brown, A. and P.W.M. Schokker (1989) Offset-printed colour charts for use with a Macintosh II Microcomputer. *ITC Journal*, **1989**(3/4), 225–228.

Dent, B.D. (1985) *Principles of thematic map design*. Reading, MA: Addison Wesley.

Kadmon, N. (1992) *An introduction to toponymy. Theory and practice of geographical names*. Pretoria: University of Pretoria Department of Geography.

The Times Atlas of the World (1992). London: Times Books.

CHAPTER 7

Statistical mapping

7.1 Statistical surveys

Statistical surveyors collect socio-economic data, either continuously, in samples or in censuses. In some countries with advanced administrative structures where the population registers of all municipalities have been converted into on-line facilities, there is a continuous updating of population data. Depending on the size of the municipality or the subdivisions discerned, other data might be collected by drawing samples from the total population and assessing the present characteristics of these samples; for example by asking those sampled where and how they spent their holidays last year, or the amount of money they borrowed in the form of loans or their current employment situation. This kind of information might be more detailed than the general data currently collected in a population register (e.g. name, gender, place and date of birth, nationality, marital status, children and their birth dates, nationality or the fact whether they own or rent their accommodation. However, although it is more detailed from a thematic point of view, it is less detailed in a spatial context, because of its sample character it might only be correct to show the data on a higher aggregation level as it would not be representative otherwise.

Most countries still have censuses every 10 years. In these censuses the population characteristics are collected for as high a percentage of the population as possible, by enumeration officers who try to check the forms when collecting them in order to have as standardized an interpretation of the questions as possible. As well as these population censuses, in many countries housing censuses are carried out, which are meant to gather data on the situation (quality and quantity/capacity) of residential buildings.

As well as population and housing censuses there are agricultural censuses (every year or every 5 years), manufacturing industry censuses and company censuses. The scale of the production, the size of the work-force, the turnover or added value or the value of the production assets will be assessed for these censuses.

These statistical data, in whichever way they have been collected, are subject to privacy regulations. If people could not be sure of privacy, the information would not be provided by them. This means that when the data are to be rendered or published in alphanumerical form they are combined first, so that information on individual, identifiable households, farms, plants or companies cannot be worked out from the data. Another method of safeguarding privacy when data are collected for 1×1 km cells, is to randomly add numbers ($+1$, $+0$, -1) to figures for individual cells, and to suppress data cells with eight or less households or less than 25 people (see HMSO, *People in Britain*, 1980) in a data grid cell, so that it is still impossible for the other households living in such a cell to work out what the characteristics of the household would be. This secrecy is required by the data suppliers, who have given the information in confidence, but has important consequences for the data users: data are combined, and may or may not be averaged, resulting in either overall figures or average figures. Visualized, this generally leads to choropleth maps or proportional symbol maps. Their characteristics will be commented upon in Section 7.5, but at this stage we should point out some visualization aspects that are directly related to the data gathering techniques.

There is a false suggestion of homogeneity in choropleth maps: values for a phenomenon like population density or the number of general practitioners per 10 000 households are represented in such a way that a suggestion of homogeneity is conveyed: values seem the same throughout the enumeration unit. Look, for example, at Figure 7.1(a): each dot represents the income of a farmer located at the site of a dot. Incomes tend to vary considerably in this example because of the variation in soils and relief, i.e. gently sloping valley farms on good soils versus small patched hillside farms on steep slopes and poor soils. When the statistical enumeration officers pass by they will collect the income data (or they might obtain them from the inland revenue), which will then be combined (because of privacy regulations) and represented as a choropleth: for every enumeration area the combined income data will be divided by the number of farms, and these average data categorized (see the choropleth method procedure in Figure 7.20). The resulting image is shown in Figure 7.1(b), represented with proportional grey values. So the extreme local differences in characteristics that geographers would be interested in, have been obliterated. This is because the enumeration area boundaries are seldom drawn with a specific phenomenon in mind – they are not relevant for this phenomenon. It would be preferable to base the boundaries on the original data (as found in Figure 7.1a), delineate the areas with similar values, and determine averages within these new boundaries, so that the choropleth result would be as in Figure 7.1(c). But cartographers are seldom in a position to do this, as the original data are not available to them. However, one should still be aware of this problematical aspect of rendering statistical data.

Another aspect is that when non-area related ratios are being represented, the impression of the choropleth will be determined more by the size of the enumeration unit than by the actual values expressed by the grey value tints: it is the size of the areas that makes the ratios represented in them stand out, while it should be the size of the original numbers they refer to. Consider the example shown in Figure 7.2. When mapped, the large marginal areas in the west and the north stand out, suggesting it is there that the number of GPs is highest. However, when one analyses the data, it is found that the largest number of GPs can be found in the darkest, urban areas (see also the example on absolute and relative unemployment in Figure 3.11).

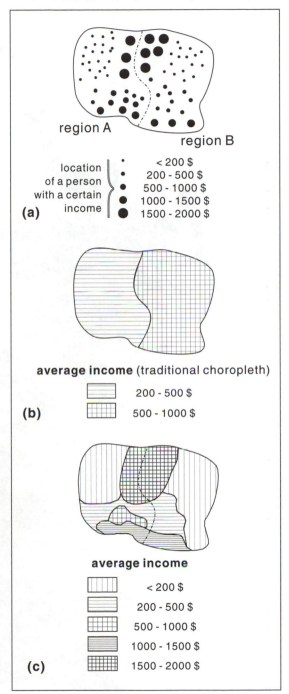

Figure 7.1 Farm income figures: (a) reality; (b) represented in the usual choropleth manner; (c) with boundaries adjusted to the phenomenon

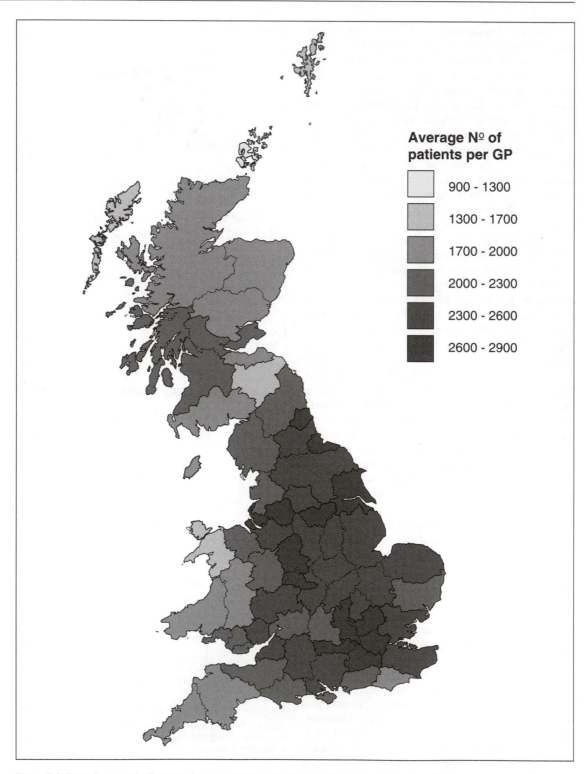

Figure 7.2 The influence of differences in enumeration area sizes on data perception in choropleths

What people see when confronted with proportional symbols related to areas is a ratio between the size of the symbol and the size of the enumeration area it refers to. When areas are compared, one actually compares these ratios, and the ratios are considered to stand for the whole enumeration area, again suggesting a homogeneity that does not exist.

In fact this problem is only offset when the original data coordinates are known (such as the addresses of the households being surveyed for the Census) so that the data can be represented within regular cells, as in the atlas *People in Britain* (1980), for example (Figure 4.11).

7.2 Data analysis

When statistical data are presented for representation, a number of questions need be asked in order to assess how useful they will be. The answers might be directly available in the metafiles that accompany the relevant file, but it is nevertheless necessary that one realizes the implications of these answers. The questions would revolve around when the data were collected, in which way, for what purpose, for which period of time and to what area they refer to, etc. These answers will indicate the utility, reliability and accuracy of the data. They will indicate whether the data have been collected in a way comparable to similar gathering exercises in the past, so that time series can be realized. The conclusion might be that, as the data were collected with a specific objective in mind, they might be biased.

After their assessment of the validity of the data, it is the data characteristics that need to be determined prior to further analysis/processing. In the first place there is the nature of the objects the data refer to: are these point-like objects, linear objects, areal objects (two-dimensional) or volumetric objects (three dimensional)? Examples are signposts, roads, fields and mountains.

The next characteristic is the type of change in the data: is this change gradual or not, and do changes occur abruptly from one place to the next? The smoothness of the change is related to distributions being continuous or discrete/discontinuous. A continuous distribution describes data that can be measured anywhere, such as air pressure or temperature. Discontinuously distributed phenomena can only be

ascertained at particular locations and not elsewhere, such as landcover or vegetation types.

The measurement scale on which the object attributes can be measured comes next. The representation of the data will later depend on this measurement scale, which is why it is important to find out about it in advance (if specific map-related measurement scales are not appropriate it might be necessary to change the measurement scale of the data). If only a very simple population map is required, and population density figures are available, some relevant threshold values might be selected and the area with lower values represented as uninhabited while the area with values higher than this threshold can be shown as the ecumene – the human-inhabited world.

The measurement scales are traditionally divided into *nominal*, *ordinal*, *interval* and *ratio* scales. Phenomena attributes are measured on a *nominal* scale when differences in data are only of a qualitative nature. Examples are differences in gender, language, religion, land use or geology.

The position on an *ordinal* scale of object attributes is determined, when it is known that their values are more or less than another, without knowing the exact distance on the scale between them, or when these concepts have a vague/inaccurate meaning, such as 'cool–tepid–hot' or 'small–medium–large'. Tepid is warmer than cool and hot is warmer than tepid; how much warmer is unknown, but it is at least possible to establish a hierarchy.

Both a hierarchy and the exact distance on the scale is available for measurements on an *interval* scale, with the restriction that it would not be possible to work out relationships/ratios between the measurements other than their distances: 8 °C is not twice as hot as 4 °C, it is 4 °C warmer than 4 °C! Likewise, 2 °C is not twice −2 °C, though it again is 4 °C warmer. This is because a random temperature value has been selected as the 0 °C point, in this case the temperature at which water will freeze when the temperature goes down.

If two locations, A and B, along the Dutch–Belgian border have heights of 4 and 8 m respectively above the Netherlands datum (Normaal Amsterdams Peil, NAP), this does not mean that the latter is twice as high as the former. It is only the difference in height between the two points that can be assessed, i.e. 4 m. The datum for heights used in Belgium is 2 m lower, so the heights of these two points on the border would be 6 and 10 m respectively in Belgium (Figure 7.3)!

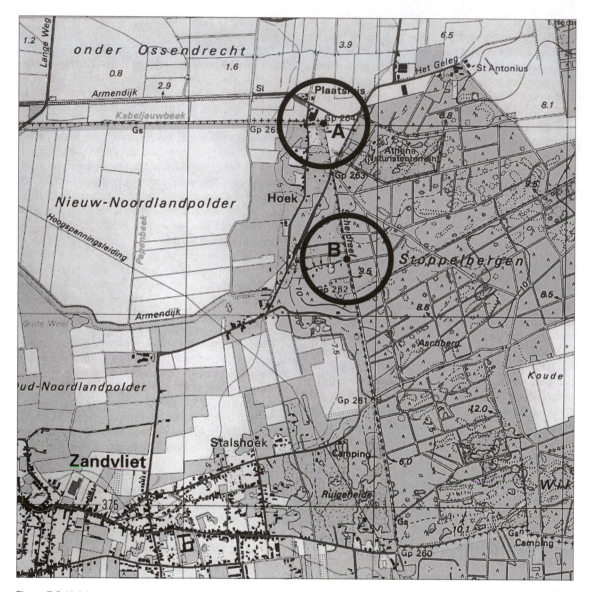

Figure 7.3 Height representation on topographic maps of the Dutch–Belgian border (topographic map 1:25 000, sheet 49G – Courtesy Topografische Dienst Nederland)

When *ratios* can be expressed, one refers to data that can be measured on a ratio measurement scale. In 1992 the Gross National Product per capita in Spain was twice that of Portugal so these are ratio data.

There is a difference between absolute and relative ratio data. Absolute ratio data are the result of direct measurements or additions of units. Examples are one's income in money (a specific number of pounds or dollars), or the number of children in a family. In the first example, income can have all values in between, with decimals, while in the second case it can be only discrete values; in other cases there might be positive and negative values, as when immigration and emigration data are compared.

Relative ratio data are absolute ratio data related to other data sets, and it is these relative data that often tell us more than absolute data, because they have been put into context. Whether a gross income figure of an equivalent of $30 000 a year is high or not will depend on one's perspective, and on what it will buy. In India (in 1995) it would be the income of a top manager, providing him with a large estate and many servants; a similar income in Alaska might only attract a primary school teacher. So one should provide some yardsticks and compare the income to the average world income or the average income in developing countries or the average income of managers. Even if it were considered low now, when compared to the remuneration for similar jobs in India in the past, it might seem enormous. So the relation to a point in the past should be established, and one would be using index values to express these relationships. For example, if in 1910 a certain job paid $800 a year, but now pays $30 000, there would be a 37.5 times increase. If the value in 1910 equated to 100, the present index value would be 3750. Expressing data in index values makes them easier to digest. As well as the index values that can be used in time series, other relative or derived values can be used in order to give more meaning to absolute data sets: examples are averages, ratios, densities and potentials.

Densities originally referred to the ratio between the population of an area and the resources available to that population. Population density values (number of inhabitants per areal unit, such as square kilometres) originate in the Malthusian era, higher densities referring to smaller means for the population to provide for itself. This concept has lost its economic relevance, but retained a relevance as an indicator for well-being/welfare, as high population density figures refer to crowded situations. To retain some relevance as a population/resources ratio, the density concept has been refined and can refer to the number of people in relation to their residential area (residential density) or the number of people in relation to the agricultural area they are cultivating (agricultural density).

Non-area-related ratios can express the relation between any two data sets; for instance, between the total number of inhabitants and the total number of general practitioners, or the relation between two subsets of the total population, e.g. between the number of members of the armed forces and the number of those in the teaching profession.

An absolute number of influenza patients of 30 000 in the Netherlands as compared to 28 000 in Luxemburg would not be reason for much comment. But if one were to convert these absolute figures into ratios, relating them to the total population numbers of the respective countries, this would result in much more information data: in the Netherlands (15 million inhabitants) only 2 in every 1000 people would be ill, whereas Luxemburg (300 000 inhabitants) 93 in every 1000 would be smitten, the influenza thus being a real epidemic. Expressed in percentages (realized by multiplying these figures by 100) these values would be 0.2% and 9.3% respectively.

Averages attempt to characterize the data sets they refer to by one number – and can only succeed in doing so if the variation in measurement is not too high. An average income of $3000 per year is a meaningless figure in a country where 90% of the population has an income of around $500 per year and 10% of the population earns over $25 000 per year. In this case, the average cited, $3000, is the arithmetic mean. A better or more useful number to describe economic conditions would have been the mode, i.e. the class in which the highest number of inhabitants would fall, e.,g. the $400–600 income class. In other cases the median value – the value above or below which 50% of the inhabitants would score with their income – could be more appropriate. Other derived measurements can be found in statistical manuals. They would also serve to describe distribution patterns of point locations, the topological or hierarchical characteristics of line patterns, and the shape of areal patterns.

One of the descriptors of dot patterns is the nearest neighbour index. This compares random patterns and actual patterns on the basis of the distance between each point location and the point location nearest to it. These index values can range between 0, when all observations are concentrated in one single point, via 1 (when the distribution is a completely random one), to 2.15, when the distribution of the points is completely regular, all distances between them being equal.

Take the province of Drenthe, Netherlands. From a map of population centres with over 10 000 inhabitants (Figure 7.4), one might get the impression that settlements are more regularly spaced in Drenthe as compared to other provinces. This might be the result of the distortion in the potential regular service city network that could have emerged, caused by the lower courses of the rivers Rhine, Scheldt and Meuse in most of the other provinces.

Now when the distribution of points is totally random within an area, the average distance between each point and the point nearest to it can be com-

Figure 7.4 Population centres with over 10 000 inhabitants in the Netherlands

Table 1 Nearest neighbour index values of places over 10 000 inhabitants in the Netherlands

Drenthe	1.6
Overijssel	1.5
Limburg	1.2
Friesland	1.18
Noord-Holland	1.16
Noord-Brabant	1.15
Gelderland	1.08
Zeeland	1.04
Zuid-Holland	1.01
Utrecht	1.0
Groningen	0.93
Flevoland	2.1

These data might still not prove that it is the disturbing influence of the rivers which causes this variation in nearest neighbour index values, but they would serve to strengthen one's resolve to analyse the data further.

The formula for the nearest neighbour index is:
$$R_n = d_o/d_e$$
while d_o is computed from $1/2\sqrt{p}$ (see above).

puted with the formula $d_e = 1/2\sqrt{p}$, p denoting the density of the points taken into account and d_e denoting the expected mean distance between each point and its nearest neighbour. If there are seven cities with over 10 000 inhabitants in Drenthe (area 2685 km^2), the density is $7/2685 = 0.002607$ km^2, so $d_e = 1/2\sqrt{0.002607} = 9.79$ km.

The average of the actual nearest distances when measured between the centres of these seven cities, d_o, is 15.66 km. The nearest neighbour index now compares these two values: d_o/d_e, which gives $15.66/9.79 = 1.6$. This value is more regular than random as compared to that of the other provinces (see Table 1).

A concept derived from physics, describing the attraction between two masses ('bodies') is the *potential*. This attraction force is equivalent to the product of their two masses, divided by their squared distance. In a watered-down version this concept describes the virtual interaction between the inhabitants of different cities (population potential) or the expected purchases in a market (market potential), etc. The population potential at a point describes the chance that people will meet each other, or have contact. This is a chance expressed for each city by the addition of population numbers of other cities divided by their distance to the city where the potential is measured (Figure 7.5). To find the potential value between which one has to interpolate to produce the final map (Figure 7.5bIII) for each city, the influence of other cities (expressed by their number of inhabitants divided by their distance to this city) has to be added up. The influence of city A upon itself can be expressed by dividing its number of inhabitants by the radius of its geographical extent.

An example of a socio-economic data set analysis is that of statistics on nature, area and number of agricultural holdings in Indonesia (Sensus Pertanian, 1973). As can be seen in Table 2, the data made available are on the number and size of individual farms and agricultural estates/plantations. Not all of these data categories are equally important, however, and if only one map were available to give as good an impression of Indonesian agriculture as possible, choices would have to be made. Data sets related to each other, such as average farm size, would require less space than their separate representation. Individual farms in Indonesia would mostly grow the staple crop, rice, and their average size would show the agricultural density, that is the pressure on the land the individual farms would be subject to. Agricultural estates, on the other hand, are sort of extra – they do not produce food for subsistence or for the local market but raise export crops: tea, coffee, rubber, oil palms, sugar. So when asked

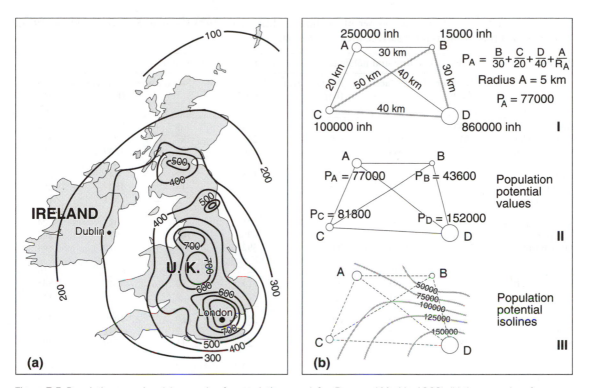

Figure 7.5 Population mapping: (a) example of a population map (after Berry and Marble, 1968); (b) the procedure for assessing the population potential

which would be most important for Indonesia's economy, the individual farms or the agricultural estates, the former would score higher. This importance can be visualized by assigning this aspect the first place in the graphical hierarchy, giving it most emphasis. That is realized by expressing it through area symbols. In addition, using proportional circles, the total agricultural acreages could be shown (Figure 7.11), allowing readers to compare the ratios between these circles and the areas they pertain to on the map. In these circles the proportion of the area in agricultural estates would increase the interpretation potential, but might make the map more complex and thereby endanger the possibilities for communication.

It is important to note that the representation in this example has actually been a data selection process (as not all the available data have been used) based on the knowledge of the phenomenon (Indonesian agriculture). That is why it is so important that those who use databases for map production have a geographical background that makes them aware of things geographically relevant.

7.2.1   Data adjustment

In order to make the data selected relevant for comparison purposes, derived figures might have to be adjusted. Fecundity figures, birth rates and death rates are examples here. Birth and death rates refer to the numbers of births or deaths per 1000 inhabitants. Originally, this concept was used to give an indication of the fecundity or salubrity of an area – but this presupposes a 'normal' population structure, i.e. a bell-shaped population pyramid. If there is an overrepresentation of inhabitants over 65 years, the death rate is bound to be higher than elsewhere, however healthy the climate might be. While in most states of the US the death rate has decreased with the heightening of the average life expectancy, it has increased in Florida, because of the enormous influx of retired people with their higher death rate. In order to give a proper image of the actual health-influencing aspects, this figure has to be corrected, for instance by assessing whether

Table 2 Agricultural surface and number of holdings in Indonesia, 1973 (Source: Sensus pertanian, 1973). Numbers of farms in thousands; surface area in 1000 ha.

Province	Individual agricultural holdings			Plantation agriculture			
	Number of farms	Surface area	Average area (ha)	Number of estates	Surface area	Average area (ha)	Total surface area
Aceh	353	373	1.06	90	263	2932	637
Sumatra Utara	816	805	0.9	265	787	2937	1593
Sumatra Barat	426	344	0.8	30	52	1349	385
Riau	199	507	2.5	135	32	243	540
Jambi	142	241	1.6	8	14	1803	255
Benkulu	84	153	1.8	7	3	511	157
Sumatra Selatan	377	703	1.8	21	21	1083	724
Lampung	446	673	1.5	38	149	3932	822
Jawa Barat	2468	1524	0.6	392	319	815	1844
Jakarta	20	19	0.9	—	—	—	10
Jawa Tengah	2765	1753	0.6	123	92	748	1845
Yogyakarta	343	181	0.57	5	5	1114	186
Jawa Timur	3066	2026	0.6	253	260	1029	2286
Bali	305	266	0.8	8	2	342	269
Nusa Teng. B	281	289	1.0	21	11	539	300
Nusa Teng. T	365	652	1.8	12	2	176	655
Timor Timur	—	—	—	—	—	—	—
Kalimantan Bar	273	981	3.6	38	14	394	966
Kalimantan Ten	100	524	5.2	11	5	524	529
Kalimantan Tim	57	92	1.6	5	2	477	94
Kalimantan Sel	257	269	1.0	31	36	1166	305
Sulawesi Sel.	648	737	1.1	92	106	1153	843
Sulawesi Tengah	132	283	2.1	12	1	118	284
Sulawesi Utara	217	351	1.6	102	16	157	367
Sul. Tenggara	102	151	1.5	30	3	118	154
Maluku	119	259	2.8	72	30	420	290
Irian Jaya	—	—	—	—	—	—	—

for a specific age group the death rate is higher or lower than in other areas. The fertility rate (the number of births per 1000 women of child-bearing age) is such a refinement over the birth rate, as it takes only the relevant age and gender group into account.

The same applies for figures in physical geography, like temperature or vegetation. When one would like to assess the effects of geographical latitude on vegetation or climate, one should try to minimize the distorting effects of relief (height, slope and aspect). For temperature, there is an easy rule of thumb which says that every 100 m shows a decrease of 1 °C, so that it is possible to reduce all temperatures measured to sea level, and thus make all values com-

parable. Although often other climate effects, such as precipitation, are also related to elevation differences, the relationships are not always so simple and linear as is the case for temperature.

The above examples illustrate the importance of data normalization before visualizing them.

7.3 Data classification

The risk of the mapping of unprocessed data resulting in unclear visualization is quite high. It is good cartographic practice to conveniently arrange the data before displaying them. This process is called classification. It can be described as systematically grouping data based on one or more characteristics. Classification will result in a clearer map image, even if it is a generalized image. To be able to classify data, one needs to know what types of data are available, which requires an operational definition of the data. If one is interested in the number of ships that entered the port of Rotterdam in 1995, it is first required to define what a ship is. Setting these definitions is not easy. Practice proved that, for instance with the classification of topographic objects, many different descriptions are in use for the same object (such as lamp-post and lamp standard). Even if one agrees on the description it is still possible that the description might not be completely clear in practice. Clearly the Golden Gate Bridge fits in the object class 'bridge', but is this also true for a plank over a small stream? When all elements adhere to the same criteria (discrete phenomena) one can count class elements. For instance, one can count the number of oil tankers entering the port. It is possible to count according to quantitative or qualitative rules, such as 10 tankers and 5 container ships, or 15 ships. When one deals with continuous phenomena such as precipitation, the data can be measured: the characteristics of the objects, rather than the objects themselves, can be quantified according to a specific measurement scale, and mapped. Four measurement scales exist, as was mentioned in the previous section.

Classification is helpful to enhance insight in the data. However, to make sense, the number of classes should be limited. Research has revealed that humans can handle up to a maximum of seven classes to get an overview and understanding of the theme mapped, at a single glance (see also Figure 7.13). The exact number of classes chosen is influenced by the type of symbolization used, the theme's geographic distribution, and the data range (the ratio between the maximum and the minimum data value).

Not everyone is convinced that classification is needed at all times. The American cartographer Tobler (1973) is of the opinion that it is unnecessary to classify statistical data. The major advantage of not classifying the data is that the resulting image is not generalized. For the legends of such maps he suggests using a continuous grey scale. In Figure 7.21 the maps on the right show the population density in Maastricht municipality mapped as a traditional choropleth map. The map on the left shows the same theme, now represented by an unclassified chor-

opleth map. The grey value of each neighbourhood in the map on the left is derived from the class to which the neighbourhood belongs according to a particular classification method. The general appearance of the maps is different. Extreme values are much more isolated in the unclassified map. However, some cartographers oppose this method since it is virtually impossible to perceive the differences between neighbourhoods that are further apart geographically. The same cartographers also claim that a limited number of grey shades, as in the maps on the left, improves the legibility of the map. If one has the choice between the two approaches, one should always ask first: 'What is the purpose of the map?' Is it necessary to be able to determine values for each enumeration area, or is it just an overview one is interested in? It should be mentioned that not many software packages offer the possibility to create unclassified maps. If one remembers Figure 4.6, it should also be noted that most GIS software packages allow one to access the database even during display, and that the user can access the values of each individual geographic unit by pointing at the unit on the map.

If one decides to classify, this process is executed according to the nature of the data. Nominal data are categorized according to taxonomic principles of the discipline involved, such as soil types, climatological zones or geological periods. The map type used to display nominal data is the chorochromatic or mosaic map (see Section 7.5.1). Ordinal data are also based on classifications defined by individual disciplines. Examples are meteorology (temperature: cold, mild, warm) and environmental sciences (forest conditions due to air pollution: healthy, normal, sick). Interval and ratio scales are both linked to quantitative data. Most census statistics belong to this category. Examples are the number of people living in a country's districts, or the percentage of children under the age of 15. Interval and ratio data are displayed on choropleth maps or isoline maps (see Sections 7.5.2 and 7.5.3). The first is primarily associated with socio-economic data collected for administrative units, while isoline maps show interpolated data derived from physical measurements such as temperature or hours of sunshine.

To reach the best possible classification several conditions have to be met, since not all classification methods are suitable for all situations. Therefore one should strive for the following:

- The final map should approach the statistical surface as closely as possible. A statistical surface is a

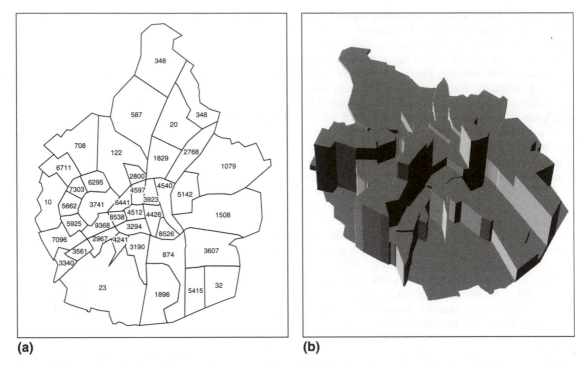

(a) **(b)**

Figure 7.6 Population densities of the neighbourhoods of Maastricht: (a) values for each neighbourhood; (b) the corresponding stepped statistical surface

three-dimensional representation of the data, in which the height is made proportional to the numerical attribute value of the data (see also Section 7.5.9). In Figure 7.6 the numbers displayed in (a) are used in (b) to give each enumeration area its height. Statistical surfaces offer the user a dramatic view of the data. When functionality (as described in Figure 5.33) is available, the user can even view the data from all possible directions (see also Section 7.5.9). Two types of statistical surfaces are distinguished: the stepped statistical surface, as displayed in Figure 7.17(c), and smooth statistical surfaces (Figure 7.17d). Stepped statistical surfaces are derived from choropleth maps, and each of the enumeration districts is clearly visible. Smooth statistical surfaces are derived from isoline maps. In this map the height at each location is defined by the intensity of the phenomenon mapped at this particular location.

The final map should display those patterns or structures that are characteristic for the mapped phenomenon. Extreme high or low values should not disappear because of the classification method.

● Each class should contain its share of the observed values.

If one adheres to these conditions the resultant map will give a clear overview of the mapped phenomenon, and it will be possible to determine values at each location in the map. A method used to adhere to the above conditions was proposed by Jenks and Coulson (1963). It can be split in three major steps.

1 Choose a map type. Since one is dealing with quantitative data it will be a choice between an isoline map and a choropleth map. In the example elaborated here, it is a choropleth map because the topic deals with population density per neighbourhood of the city of Maastricht (Figure 7.6).

2 Limit the number of classes. If one looks at the grey value of a particular geographic unit, and wants to determine the corresponding grey value in the legend, it is not possible to use more than eight classes. This seems to be contradictory to the statement made earlier which said that the map should be as close to the statistical surface as possible. This objective could induce cartographers to discern as many classes as possible, or even to

renounce from any classification at all. However, it could also be argued that if one intends to grasp all variations in the mapped phenomena in a single moment, the number of classes should be reduced.

3 Define the class limits. This is the most difficult step in the classification process. Many different methods exist, although most software packages offer only three different options. These are either to split the number of the observed values equally over all classes, to have equal class size in respect to the range of the observed values, or to define one's own classification method. This last option allows the user to apply every conceivable method. When class limits have to be defined external to the current software packages, the user should be aware of potential classification methods. The most important classification methods will be explained below. In general, one can distinguish between graphic and mathematic methods. The first type of method will not be found in the classification options of the mapping software.

7.3.1   Graphic approach

Break points

When all observed values are sorted in ascending order and subsequently plotted in a graph, such as the one in Figure 7.7, it is sometimes possible to observe discontinuities. In Figure 7.7, four of these discontinuities are clearly visible: between the observed values 14 and 15, 29 and 30, 38 and 39 and 40 and 41. This type of graph is, in this context, also known as an observation series. The discontinuities, called break points, can function as class boundaries since they are natural breaks in the observation series. It is obvious that not all data sets will show break points, and if they do, the chance that there are enough for the number of classes planned is quite uncertain. In the example only four class boundaries can be distinguished.

Frequency diagram

A frequency diagram can also be used in a search for discontinuities. If found (Figure 7.8a) they can function as class boundaries. A frequency diagram is especially useful when a large number of observations are involved. The chance to find discontinuities increases when the value along the horizontal axis are grouped together. Figure 7.8(b) shows an example. The existence of break points is strongly influenced by the size of the intervals applied to group the data.

Cumulative frequency diagram

In a cumulative frequency diagram, as displayed in Figure 7.8(c)I the frequency of the occurrence of the observed values is added up. Changes in orientation of the curve (see the small arrows in the diagram) indicate the break points. The values along the vertical axis of those break points are the class boundaries. Again, as with the observation series, there is a limited chance that the number of class boundaries needed will indeed be found using a graphic approach.

7.3.2   Mathematic approach

When looking at Figure 7.7, it is possible to draw a curve along the top of all the bars. The shape of this curve can be described by a function. The mathematical approach uses this type of curve to determine the nature and location of the class boundaries. In Figure 7.8(e) curves with several common functions are drawn. The first two methods discussed below can also be executed using a cumulative frequency diagram (Figure 7.8cII and III).

Equal steps

In this method the class width is equal for all classes. This method should be applied when the curve created from the observation series is linear. The classes can also be determined graphically, as can be seen in Figure 7.8(c)II. If one applies the mathematical approach, the highest value is subtracted by the low-

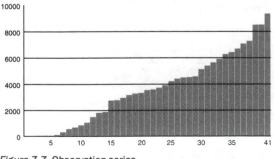

Figure 7.7 Observation series

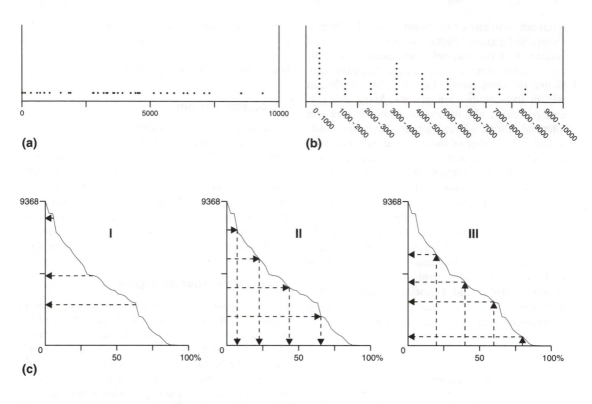

(a)

(b)

(c)

equal	quantile	arithmetic	geometric	harmonic	nested
10 - 1882	10 - 708	10 - 634	10 - 39	10 - 12	10 - 1667
1882 - 3753	708 - 2967	634 - 1881	39 - 154	12 - 17	1667 - 3585
3753 - 5625	2967 - 4241	1881 - 3753	154 - 607	17 - 25	3585 - 6067
5625 - 7496	4241 - 5925	3753 - 6248	607 - 2384	25 - 50	6067 - 9368
7496 - 9368	5925 - 9368	6248 - 9368	2384 - 9368	50 - 9368	

(d)

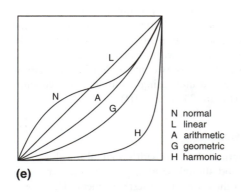

N normal
L linear
A arithmetic
G geometric
H harmonic

(e)

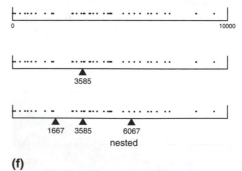

3585

1667 3585 6067
nested

(f)

Figure 7.8 Different classification methods

est value, and the result is divided by the number of classes $((9368 - 10)/5 = 1872)$. The result is used as a constant C in the following formula to determine the class boundaries:

lowest value $+ C + C + C + C + C =$ highest value

The table in Figure 7.8(d) gives the boundary values, while the stepped statistical surface and the corresponding choropleth map can be found in Figures 7.10(a) and 7.9(a) respectively.

Quantiles

This method splits the number of observations proportionally over the number of classes chosen. The name of this method is adapted according to the number of classes: when applied to four classes it is called quartiles; with five classes, quintiles, etc. This method can be advised when the areas of the geographic units have a comparable size, and when one is interested in correlation between different characteristics of the units. In the Maastricht example the 41 neighbourhoods are split over five classes. The lowest class has nine observed values. The results of the method are found in the table in Figure 7.8(d). The corresponding stepped statistical surface is found in Figure 7.10(b), while the resulting choropleth map is shown in Figure 7.9(b). Figure 7.8(c)III shows how the class boundaries can also be derived from the cumulative frequency diagram.

Arithmetic series

An arithmetic series is a series of numbers in which each term can be directly derived from the previous term by adding a constant increase. The curve labelled A in Figure 7.8(e) belongs to this method. The class boundaries can be calculated according to the formula below, assuming one is interested in having five classes:

lowest value $+ C + 2C + 3C + 4C + 5C =$ highest value

C has a value of 624, and is calculated via: highest value minus lowest value divided by the number of constants C in the formula $((9368 - 10)/15)$. The table in Figure 7.8(d) shows the class boundaries, and Figures 7.9(c) and 7.10(c) show the stepped statistical surface and choropleth respectively.

Geometric series

For geometric classification, the curve labelled G in Figure 7.8(e) is derived from a geometric series. Each following term is derived from the previous term by multiplying it with a constant C, the ratio of the series. To determine the class boundary values according to this method, the logarithm of the highest value and of the lowest value have to be determined. These values are then subtracted from each other and divided by the number of classes, resulting in the logarithm of the constant C, which can be computed as follows: $(\log 9368 - \log 10)/5 = 0.594$. C is used in the formulae:

log highest value $- C =$ log second highest value
log second highest value $- C =$ log third highest value, and so on

From the results the antilogarithm is defined, resulting in class boundaries as given in the table in Figure 7.8(d). The stepped statistical surface which belongs to this method can be seen in Figure 7.9(d) and the choropleth map in Figure 7.10(d).

Harmonic series

For a harmonic classification, a series is defined harmonic when the reciprocal values of the terms can be defined as an arithmetic series. The curve labelled H in Figure 7.8(e) is linked to this method. Class boundaries are defined when one calculates the difference between the reciprocal values of the highest and lowest value, and then divides this result by the number of classes $((1/9368 - 1/10)/5 = 0.01998)$. This results in the series ratio C. A formula similar to the one applied to calculate the class boundaries according to geometric series is executed:

reciprocal highest value $- C =$ (reciprocal highest value $- C) - C =$ ((reciprocal highest value $- C) - C) - C$, and so on

From the resulting values the inverse is calculated and used as class boundaries. (See the table in Figure 7.8(d) for the exact values.) The method results in special attention for the low values in the observation series as can be seen in the table. The map in Figure 7.9(e) and the stepped statistical surface in Figure 7.10(e) confirm this.

Nested means

To determine class boundaries with this method, first calculate the average of all the observed values. In the

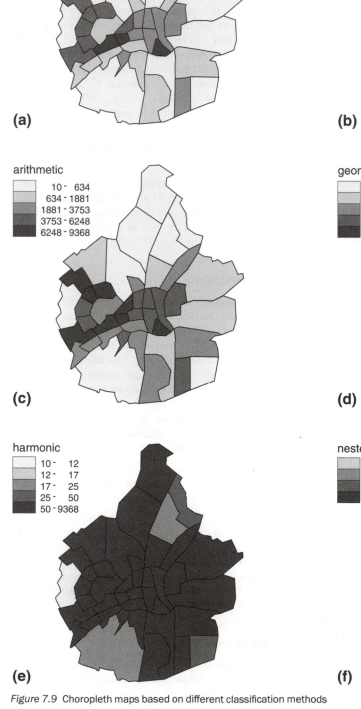
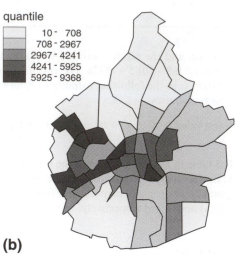
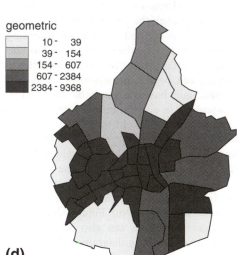
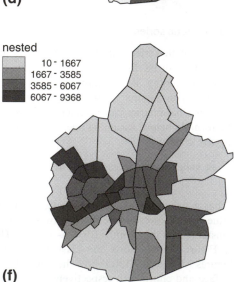

Figure 7.9 Choropleth maps based on different classification methods

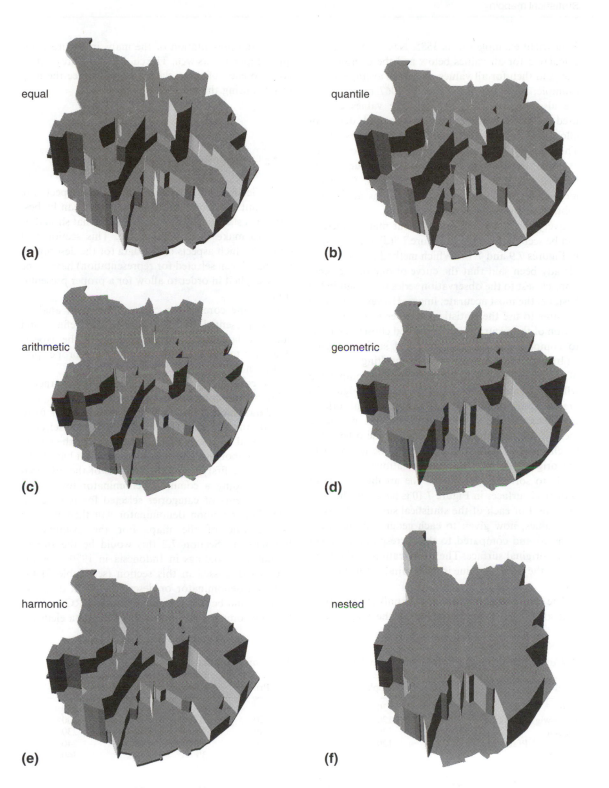

equal (a)

quantile (b)

arithmetic (c)

geometric (d)

harmonic (e)

nested (f)

Figure 7.10 Stepped statistical surfaces

Maastricht example this is 3585. Next, the average is calculated for all values below and the overall average, and then for all values above the average (in the example, these will be 1667 and 6067 respectively); see also Figure 7.8(f). These three values can be used as class boundaries. This method does not allow one to choose the number of classes freely. There always has to be a multiple of two. In the example, the method is used to define four classes, as is shown in the table in Figure 7.8(d). A derived method works with standard deviations added and subtracted from the averages.

Every method results in different map images, as can be seen in the table in Figure 7.8(d) and the maps in Figures 7.9 and 7.10. Which method is best? It has already been said that the curve of one of the functions closest to the observation series will result in the best, i.e. the most accurate, image. However, it is also possible to use the statistical surfaces to get an indication of the accuracy of the method chosen. One has to compare the original statistical surface (Figure 7.6b) with the statistical surface resulting from one of the classification methods. A graphic comparison will immediately show that some maps in Figure 7.9 are not at all similar to the original statistical surface. For instance, the nested means and the geometric method are obviously 'wrong'. For some others it is much more difficult to judge which one is closest to the original surface. Some calculations have to be made to solve this problem. The attribute value of statistical surfaces in Figure 7.10 is based on the class average. For each of the statistical surfaces the average values, now given to each geographic unit, are summed and compared to the corresponding values of the original surface. The classification method closest to the original value is most suitable for this data set.

The class boundaries are not the only characteristic that affect the image perceived by the map reader.

The visual presentation of the maps in Figure 7.9 is quite important as well. The choice of the grey values given to each of these classes will influence the map reader during the map-reading process.

7.4 Cartographical data analysis

From Section 7.2 it is clear which data aspects are important for communication, how they can be best described statistically or otherwise, or what should be done to make them comparable. This section will describe which aspects of the data (or the descriptors that have been selected for representation) have to be made explicit in order to allow for a proper presentation.

First, the core of the cartographic data analysis will be presented, and this core will be refined and added to later on, in order to show all the aspects that have to be taken into account for a proper presentation.

The core then, consists of assessing the characteristics of the components of the information, and deciding which graphic variables (see Chapter 6) to use for them. Using the graphic variables selected, one should be able to convey to the reader the nature of the components and the differences that have to be shown. The first step in the analysis of the information is finding a common denominator for all the data elements or categories selected for representation. This common denominator will then be used as the title of the map. For the example of Indonesia in Section 7.2 this would be the size of agricultural holdings in Indonesia in 1970. For the data set discussed in this section (see Table 3) the common denominator or descriptor of all data elements would be fruit production in the FRG in 1967. It is the common denominators of the data elements

Table 3 Fruit production in Germany in 1967 (source: Westerman Schulatlas, 1970)

Land	Apples	Pears	Prunes	Cherries	Other	Total
Saar	30	10	20	10	10	80
Schleswig Holstein	120	20	20	20	20	200
Hessen	130	20	40	30	10	230
Rheinland-Pfalz	120	30	40	40	10	240
Bavaria	210	40	50	30	10	340
Nordrhein-Westphalia	280	60	40	40	20	440
Lower Saxony	390	40	40	30	10	510
Baden-Württemberg	900	160	100	30	20	1210

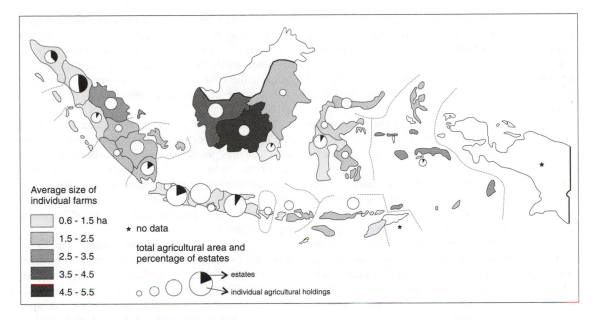

Figure 7.11 Agricultural holdings in Indonesia, 1970

that will be used later on for identifying the data set, through their application as map titles.

Then, as a next step, the data variables or components, i.e. those aspects of the attributes that vary or change from data element to data element, should be assessed and their character described. For Indonesia (see Table 2, Figure 7.11) these information components would be area used for agriculture, proportion of the agricultural area in estates, average farm size and a geographical component (the provinces). For the soil map in Figure 3.5 the components would be the geographical location of the sample site and the various soil units discerned. Then one has to determine the measurement scale at which the attributes in these components are measured. If this has already been settled during the data analysis phase described in Section 7.2, so much the better, but as the data aspects one wants to render might change during this visualization operation, and data aspects might have to be redefined in order to answer reformulated communication objectives, these measurement scales might change as well. Anyway, the measurement scales of the components will be nominal, ordinal, interval or ratio scales, or they might be geographical.

In the Indonesia example (Table 2) there are four components:

Area agriculturally used	ratio, absolute
Percentage in estates	ratio, relative
Average size of farms	ratio, relative
Provinces	geographical

The fruit production example (Table 3) has three components:

Production size	ratio, absolute
Fruit type	nominal
Länder	geographical

The soil information example (Figure 3.5) has three components:

Soil type	nominal
Groundwater category	ordinal
Soil sample locations	geographical

It is important to assess not only the nature of the components, but also their length and range. The *length* of the components refers to the number of classes or categories that will be discerned. In the Indonesia example there are six farm size classes; in the fruit production example one discerns five fruit types; and the soil map example has about 15 soil types.

As well as the length, the *range* of the data (if this attribute aspect is measured on an interval or ratio scale) should be assessed. In the Indonesia example the total agricultural area per province ranges between 190 and 22 860 km². The average farm size range is between 0.5 and 3.6 ha. In the fruit production example it is between 85 000 and 1 220 000 tons (a ratio of 1:14) per Land and between 10 000 and 900 000 tons (a ratio of 1:90) for individual fruit types.

Assessing the length and range of the components provides essential information as there is a maximum range of values that can be visualized given a specific representation method.

As indicated in Chapter 6, the maximum number of colours to be discerned on a map, while still providing an overview and the possibility to visually isolate each category, is 8. This number might be extended a bit by adding other graphic variables, such as shape (regular patterns). As can be seen from Figure 7.12, comparing proportional circles allows one to have a range between 1 and 2500 (so the largest circle to be discerned can be 2500 times larger than the smallest circle to be discerned, i.e. being different from a dot). If shown as a pie graph, the relation between the smallest sector in

the smallest circle and the largest sector in the largest circle would be 1:275. Larger differences could not really be visualized through the pie graph method. If other methods were used, e.g. composite bar graphs, the ratio in overall sizes could not exceed 1:100, while that of subdivisions could not exceed 1:10.

The final aspect of the cartographic analysis of the data to be portrayed is the information hierarchy: one has to determine what aspects are most important, what are least important, and what data categories come in between in which order. This has to be determined as this information hierarchy has to be translated into a graphical hierarchy (see Chapter 6).

Now all the building blocks would seem to be available to match the data components to the graphic variables that convey an idea of the kind of differences expressed by the measurement scales: qualitative differences by rendering nominal components, ordered differences for rendering ordinal components, distance differences for rendering interval components and proportional differences for ren-

name	shape	max. range	total quanti-ties	inner struc-ture	partial compa-rison	between parts and total	individual values		overall values	
							esti-mating	measu-ring	esti-mating	measu-ring
pie chart	 	a) 1:275 b) 1:2500	excel-lent	fair	possible	fair	hard	hard	limited	possible
pie chart with different scales	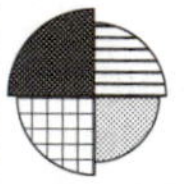	a) 1:140 b) 1:1400	hard	fair	fair	bad	fair	fair	very hard	hard
bar graph		a) 1:5 b) 1:100	fair	fair	fair	fair	fair	fair	fair	fair
histo-gram	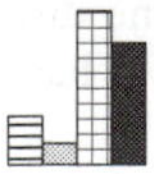	a) 1:10 b) 1:100	hard	excel-lent	fair	hard	fair	fair	hard	hard

Figure 7.12 Effectiveness of different diagrams (after Gächter, 1969)

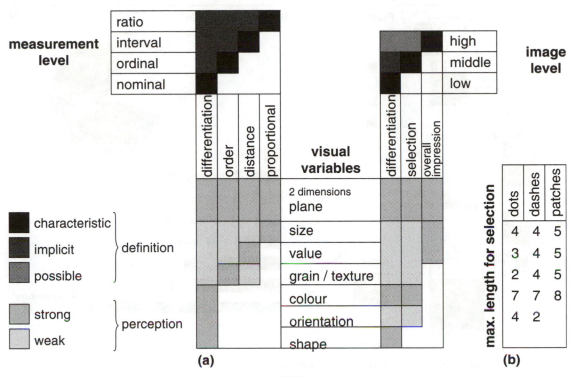

Figure 7.13 Cartographic information analysis (from Geels, 1987)

dering ratio components (Figure 7.13; this is in fact an extension of Figure 6.7).

But there may be some additional considerations before the graphic variables can be applied to render data characteristics according to one of the mapping methods discussed under Section 7.5. After the relation of the data to the earth's surface and their measurement level has been assessed, and its quality evaluated and locational aspects (shapes and topology) have been taken account of, a preliminary visualization of the data will be effectuated. This preliminary visualization has an analytical function: it will show trends, patterns, etc., to hang on to during further transformations, which might be necessary. Such transformations may be required depending on the communication objectives as well as the audience: more generalized, simpler and less abstract maps need to be produced for a less schooled audience. The results of the data evaluation might be taken into account, requiring a scale reduction, for example, or an aggregation of the data. Only after all these steps have been gone through will the final map answer the requirements.

Of course, during the data analysis process, one should ask oneself whether it is essential to map the data, and whether it might not be more profitable to graphically portray them otherwise. If the spatial aspects of the data are well-known, they may be visualized otherwise, in order to allow for better comparisons. Figure 7.14 provides some examples. The fruit harvest in the Federal Republic of Germany in 1967 is portrayed in Figure 7.14(a) (see also Table 3): the squares in each region are proportional to its production, and are subdivided for the various fruit types (apples, pears, cherries, etc.). From this map it is possible to get a good impression of the differences in the overall fruit production in the FRG, but it is very difficult to compare, for instance, the pear production in Baden-Würtemberg to that in Bavaria or Nordrhein-Westfalen. If that were one of the aims of the visualization, then another type of visualization would be called for, and Figure 7.14(b) would be the answer. Here, because of the standardized width of the bars for each region it is easy to compare the production for each fruit type. So the aims of the visualization will also influence the choice of the graphical representation method.

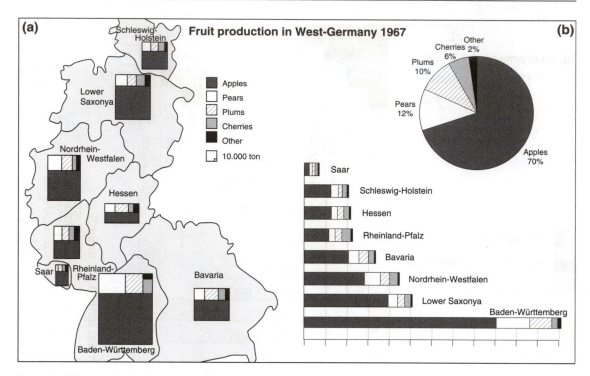

Figure 7.14 Fruit production in West Germany, 1967 (Courtesy Westerman)

7.5   Mapping methods

Mapping methods are standardized ways of applying the graphic variables for rendering information components. In these methods not only the measurement scale is taken into account, but also the nature of the distribution of the objects (whether they refer to points, lines, areas or volumes), whether their distribution is continuous or discontinuous, and whether their boundaries are smooth or not.

The result of a specific combination of graphic variables according to such a standardized method is called a map type. Around nine important mapping methods and nine resulting map types can be discerned. These methods and their results will be described in this section as well as the specific problems in constructing them and the procedures for their production, the transformation possibilities to and from other map types and some general issues in their interpretation. Figure 7.15 shows the various map types that will be discerned here in relation to each other. The subdivision is based on characteris-

tics of the objects rendered ((dis)continuity and spatial reference) as well as their attributes: measurement scale and corresponding graphical variables.

As was mentioned in Section 7.4, it might be necessary, because of reformulation of communication objectives, to select other map types than the one the data have been inventorized in. This would call for transformation of the map type; either by going back to the original data (which would be the best method), or, if these would not be available, by transformation of the source map. The arrows in Figure 7.16 show a number of these transformations: the dot map can be transformed into the choropleth map (see Section 7.5.2) through counting the dots per enumeration area, dividing the number by the area's surface, and expressing this ratio through grey tints (Figure 7.20). The same dot map could be transformed into a chorochromatic map (Section 7.5.1) through setting a threshold value and treating this in a binary way (all enumeration areas above this value have value A, all others value B). Again, the dot map could be turned into an isoline map (Section 7.5.3) by moving a transparant template with a circle

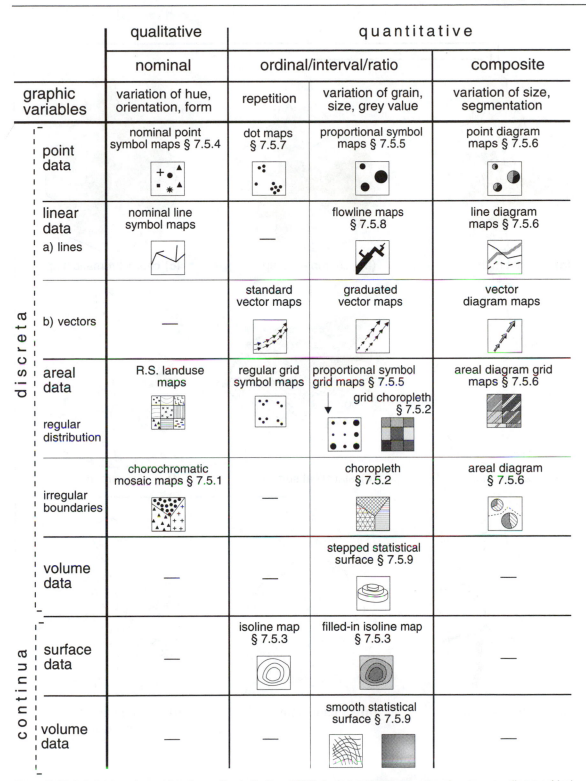

graphic variables	qualitative	quantitative		
	nominal	ordinal/interval/ratio		composite
	variation of hue, orientation, form	repetition	variation of grain, size, grey value	variation of size, segmentation
point data	nominal point symbol maps § 7.5.4	dot maps § 7.5.7	proportional symbol maps § 7.5.5	point diagram maps § 7.5.6
linear data a) lines	nominal line symbol maps	—	flowline maps § 7.5.8	line diagram maps § 7.5.6
b) vectors	—	standard vector maps	graduated vector maps	vector diagram maps
areal data regular distribution	R.S. landuse maps	regular grid symbol maps	proportional symbol grid maps § 7.5.5 / grid choropleth § 7.5.2	areal diagram grid maps § 7.5.6
irregular boundaries	chorochromatic mosaic maps § 7.5.1	—	choropleth § 7.5.2	areal diagram § 7.5.6
volume data	—	—	stepped statistical surface § 7.5.9	—
surface data	—	isoline map § 7.5.3	filled-in isoline map § 7.5.3	—
volume data	—	—	smooth statistical surface § 7.5.9	—

(Left margin labels: *discreta* for point, linear, areal and first volume rows; *continua* for surface and volume data rows.)

Figure 7.15 Subdivision of map types (according to Freitag, 1992), based on measurement scale, corresponding graphical variables and (dis)continuity of the data

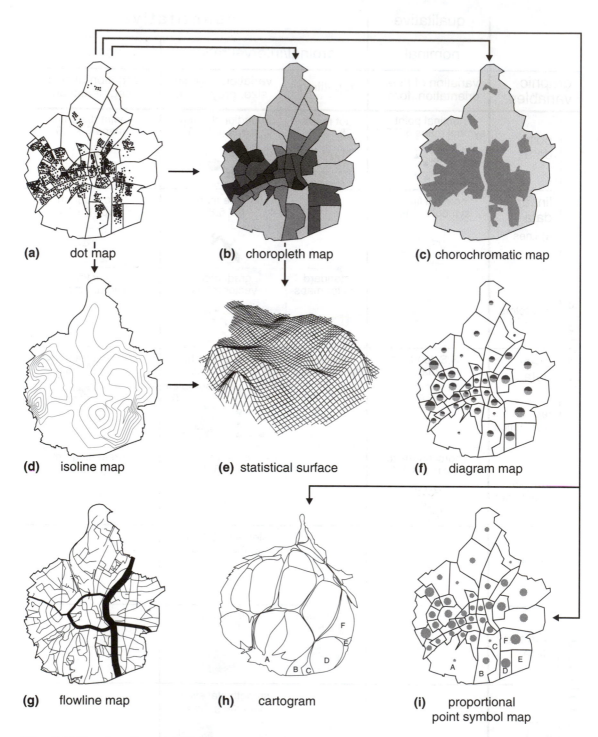

(a)   dot map

(b)   choropleth map

(c)   chorochromatic map

(d)   isoline map

(e)   statistical surface

(f)   diagram map

(g)   flowline map

(h)   cartogram

(i)   proportional
point symbol map

Figure 7.16 Transformation possibilities among maps

drawn on it over the dot map, and noting the number of dots within the circle at any location. Between the values found thus, isoline boundary values can be interpolated. When these locations are linked, the result will be an isoline map. In turn, this isoline map can be transformed into a statistical surface (i.e. draw the isolines in perspective, then draw any next isoline on a higher plane, and drape a grid over it, etc.). Another method to produce a statistical surface (Section 7.5.9) from a dot map would be to draw lines at equal distances from the dot pairs (Figure 7.17b), compute the area between these lines, and assign these areas a height proportional to the ratios between the dot value and the area between the lines. As a next step, these heights can be drawn in perspective (Dahlberg 1967; see Figure 7.17c). This statistical surface would then be used as data model, for checking whether choropleth maps (such as that in Figure 7.17c) would be a suitable rendering of the data. This stepped surface data model is the result of a specific view of the data, regarded here as phenomenon with incremental changes. Another view of the data – as a phenomenon with continuous changes – would lead to a smooth surface data model (Figure 7.17d) and corresponding isoline maps (Figure 7.17f).

In nearly all cases, these map transformations will lead to an information loss, because, when transformed back, the original situation cannot be restored. In many cases, this information loss will be taken for granted if it means that the communication objectives can be reached. More accurate data that do not get through to the map user are less useful than generalized data that do get through. Still, because of this information loss it would be advisable to go back to the original data.

7.5.1   chorochromatic maps or mosaic maps

This term has been coined by combining the Greek words for 'area' (*choros*) and 'colour' (*chroma*). So originally this method rendered nominal values for areas through different colours. But the term is also used for rendering nominal area values through black and white patterns. Both pattern (difference in shape) and differences in colour will give the map reader the impression of nominal, qualitative differences.

The most important condition for the outlook of these patterns or colours is that only different nominal qualities are being rendered, and that no suggestion of differences in hierarchy or order is being

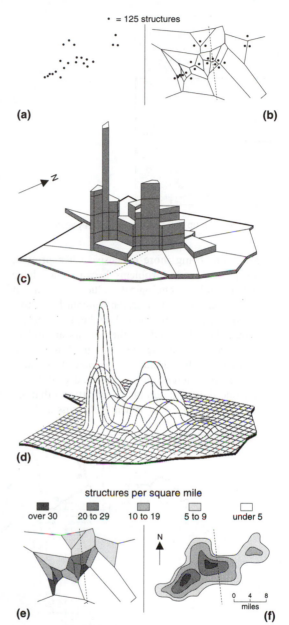

Figure 7.17 Transformation of a dot map into a statistical surface (after Dahlberg, 1967)

conveyed. This could be done easily by using different colours of black and white patterns amongst which none should have more impact (darker, brighter) than others (Figure 7.18). On the other hand, the colours or patterns selected should be easily discernible one from another.

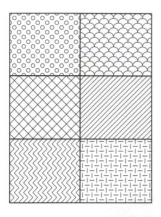

Figure 7.18 Stick-up patterns with equal grey value

Apart from being more expensive when printed, colours present extra problems because they have associative and psychological values as well; in small areas on the map or screen it might be impossible to discern a particular colour because perception would be influenced by the colours around it. Saturated colours should be only used for small areas, otherwise they dominate the image too much.

When using different patterns instead of different colours, one should preferably select patterns that are comparable in dimension. When both gross and fine patterns are applied, even if they have the same overall percentage black the more coarse patterns stand out.

When chorochromatic maps are used for the representation of non-area-related phenomena, like religion or language, the image presented to the map-reader might be influenced too much by the actual size of the areas taken by specific colours or patterns. The map-reader might assume that sizes of areas might be proportional to numbers of peoples with specific qualitative characteristics. Take Figure 7.19, for example, where dominant languages have been rendered. From this map a non-geographically schooled map-reader might assume that as they cover equally sized areas, the numbers of the English and aboriginal language communities might be equal. The high percentage of speakers of aboriginal languages in the interior refers to a very sparsely populated area, however; whereas population density is much higher along the coasts. A correction, to avoid such inaccurate first impressions, would be to add a diagram showing the actual numbers involved.

A special case of a chorochromatic map would be a grid map in which the dominant phenomenon within each grid cell would determine the assignment of a particular colour or pattern to the cell, in order to

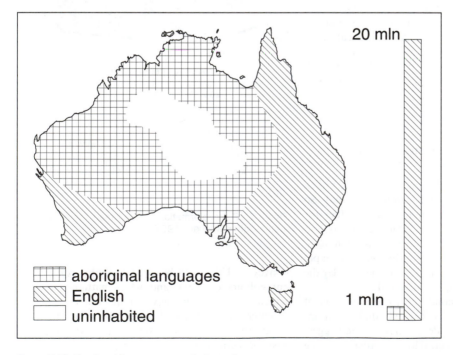

Figure 7.19 Dominant language areas in Australia

designate that phenomenon. Remote sensing imagery that has been interpreted is a type of (grid) chorochromatic map. The basis for such a map would be the images of some bands, which show the intensity of the radiation reflected from the area represented by that grid cell. A combination of these reflection values together determine – through comparison with existing maps or through taking samples from the area – what characteristic this combination of reflection values for the various bands refer to. So the actual interpretation consists of the transformation of a number of grid choropleth maps (see below) into a grid chorochromatic map.

7.5.2   Choropleth maps

The word choropleth is made up of two Greek words, *choros* for 'area' and *plethos* for 'value'. So it is values that are being rendered for areas in this method. The values are calculated for the areas, and expressed as a stepped surface, showing a series of discrete values. As these values are represented through area symbols they can only be relative values. It is the differences in grey value or in the intensity of a colour that denote differences in intensity of a phenomenon, such as differences in density. Because differences in grey values are used, a hierarchy or order between the classes distinguished can be perceived as well. But when correctly applied, percentages or densities that are twice as high are represented by a grey value that is twice as dark.

Generally the darker the grey values, the more intense or the higher the densities of the phenomenon. Another guideline is that the darker the area tints, the less favourable the conditions of the phenomenon.

It might sometimes be difficult to combine these two guiding rules. Literacy might be taken as an example: to render increasing literacy percentages on a global map through tints that increase in value would put the map reader aware of these rules on the wrong foot, as the less favourable condition would be represented by lighter tints. In cases such as these one would just change the definition of the phenomenon, using map illiteracy instead of literacy percentages, and all would be well.

There are two main types of choropleths: density maps (i.e. those that portray ratios in which the areas covered are accounted for in the denominator) and non-geographical ratio maps (e.g. the percentage of people over 65 in the total population). From a map use point of view it is important to distinguish

between these two types, as the visual impression of choropleth maps is governed by both the tint of the areas and their size. When the area is not accounted for in this ratio, this might lead to distorted images (as in the example in Figure 7.2).

The production procedure for both types of choropleth map is shown in Figure 7.20. The starting point will be absolute values for a specific phenomenon, e.g. the number of people or the number of doctors. Now in order to see whether these absolute numbers are in fact more or less than is to be expected they are put into perspective by relating them to other absolute figures, like the size of the areas these numbers have been collected for or the number of the total population. So ratios between these two sets of absolute figures will be determined. The next step will consist of categorizing all these ratios into a limited number of classes (see Section 7.3). The limiting factor in the number of classes will be the number of different grey values that can be distinguished. The maximum number of grey values is five (see Section 6.3). This range of classes can be extended by adding another colour or pattern, but with seven classes the maximum has been reached. Finally, all areas that fall into a specific category will be assigned the grey value for this category.

The aim of categorizing the ratios for a choropleth would be to improve the possibilities for communication of the information. Generally speaking, through categorization the image will be simplified, and the existence of trends or patterns will be better visualized. A condition is that differences within classes have been minimized and differences between classes maximized, that the differences between the statistical surface of the unclassified data and that of the classified data would be as small as possible, and that boundaries suggested by the classified model coincide with data boundaries in reality.

As these conditions cannot always be met, a case has been made for unclassed choropleth maps (Tobler, 1973). With present-day plotters, area patterns can be generated with a non-incremental increase in the percentage inked, and therefore in grey value, so that these percentages inked can proportionally represent the ratios that have to be mapped. Of course, this can also be effectuated for classed data, but the advantage for proportional representation for all data is that no boundaries are suggested because of areas belonging to different classes where in fact only small regional differences occur (see Figure 7.21).

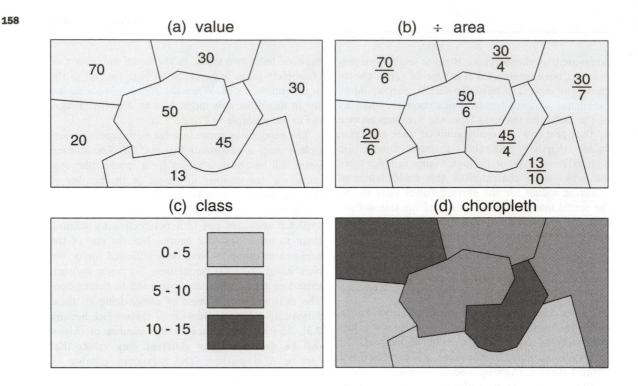

Figure 7.20 Production procedure for choropleth maps

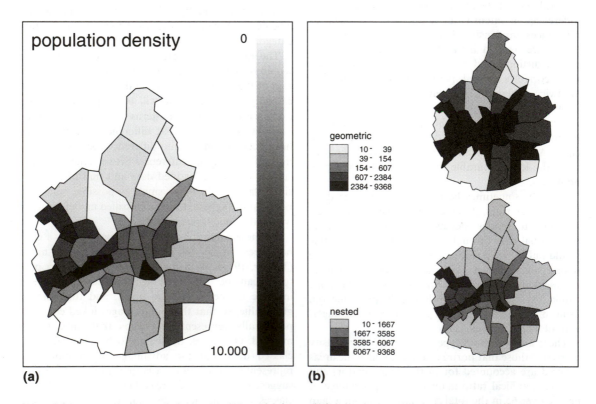

Figure 7.21 Unclassified choropleth map (a) of Maastricht's population density, compared with (b) two classified maps

Other corrections against the assumption of boundaries in locations where they do not occur in reality would be either a re-demarcation of enumeration area boundaries, or a representation in a grid-cell mode. A choropleth map with boundaries that have been adjusted to the occurrence of the phenomenon has been termed a dasymetric map. If in the dot map in Figure 7.22(a) areas with a similar density of

dots are demarcated against another, and density values are determined anew within these new boundaries, a dasymetric map would be generated (Figure 7.22f). This would be a perfect operation to effectuate on screen, but no computer software is able to perform this yet.

As can be seen, by comparing this dasymetric map or the grid-cell choropleth in Figure 7.22(b) with the

Density of student population in Utrecht

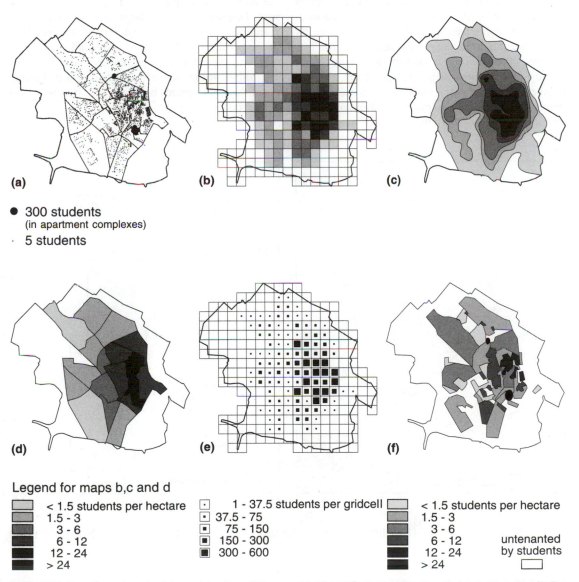

● **300 students**
  (in apartment complexes)
· **5 students**

Legend for maps b,c and d

	< 1.5 students per hectare
	1.5 - 3
	3 - 6
	6 - 12
	12 - 24
	> 24

⊡	1 - 37.5 students per gridcell
⊡	37.5 - 75
▪	75 - 150
▪	150 - 300
▪	300 - 600

	< 1.5 students per hectare
	1.5 - 3
	3 - 6
	6 - 12
	12 - 24
	> 24

untenanted
by students
☐

Figure 7.22 Transformation of a dot map into choropleth maps (b), (d) and (f), a proportional symbol map (e), and an isoline map (c)

original choropleth based on statistical enumeration areas (Figure 7.22d), the density image of the original data has already been improved upon considerably. The area with a high density protruding eastwards in Figure 7.22(d), a result of the random demarcation of the enumeration area, which was not based on the actual phenomenon at all, has been obliterated in Figures 7.22(b) and (f). The false suggestion of homogeneity of the densities within the enumeration areas has been lessened because of the small size of the grid cells.

So, a false impression is created by choropleth representation of non-area-related ratios. Two solutions offer themselves: correction of the image by adding weights or changing the base map. The first involves incorporating, as a sort of weight factor, proportional symbols denoting the actual absolute numbers of the primary data set. In a map of the representation of doctors, put into context by expressing it as a ratio of the number of doctors and the total population numbers, the number of doctors would be the primary data set, visualized by proportional symbols, against a background of a choropleth map showing the ratio of doctors to patients.

Another solution is the adaptation of the size of the enumeration areas in such a way that they are made proportional to the number of the secondary data set. In our medical example, the enumeration areas would have to be made proportional in size to the number of inhabitants or patients. In this way unfavourable ratios could be weighted against the number of people affected in this unfavourable way. Disproportionately large medical practices in marginal areas could not dominate the resulting image because the number of people affected in this way, and therefore the area rendered with this dark grey value, would be relatively small. Another example of these cartograms is Figure 7.23, in which areas are proportional to data.

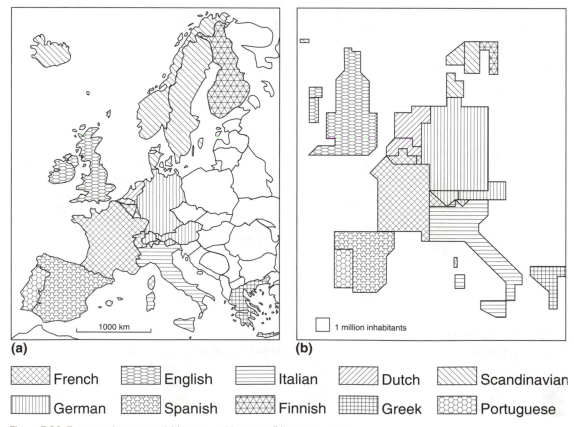

(a)   1000 km

(b)   1 million inhabitants

French   English   Italian   Dutch   Scandinavian
German   Spanish   Finnish   Greek   Portuguese

Figure 7.23 European languages; (a) in geographic space; (b) as a cartogram

Despite all these shortcomings, which require corrections or warnings to the map-readers, choropleth maps are the type of map most used for the representation of socio-economic data. This is because their construction is relatively straightforward, and can be computer-generated easily.

7.5.3 Isoline maps

Contrary to choropleth maps, which view the data set they have to represent as discrete values only valid for specific areas, isoline maps are based on the assumption that the phenomenon to be represented has a continuous distributon and smoothly changes in value in all directions of the plane as well. The Greek word *iso* means 'equal', and an isoline is a line that connects points with an equal value: equal height above sea level, equal amount of precipitation, or an equal population density. The values that serve as a starting point for the construction of isolines can be measurement data that apply either to point locations or to areas. Let us first look into the production procedure for point-data-based isolines.

In Figure 7.24(a) the location of a number of weather stations is indicated, with precipitation data (it is customary for climatological maps to use 30-year data averages). These data are now categorized into a number of classes. In the example the data, which range from 28 to 104 inches per year, have been subdivided into nine classes (<30, 30–40, 40–50, 50–60, 60–70, 70–80, 80–90, 90–100, >100). Now these class boundaries have to be drawn on the map, and this is being effectuated through interpolation. Let us take the 70-inch boundary first. There are no points in Figure 7.24(c) with the value 70, but we know such points, when constructed, would have to lie in-between data-point pairs with values on the opposite sides of 70 inches. So, to quote the map, in-between 73 and 65, for instance. Assuming the change in precipitation values occurs linearly, one would be able to pinpoint exactly the location: when the two data points with values 73 and 65 (there is a difference in value of 8 inches between them) are linked, a point with the value 70 (3 inches less than the 73-inch point) would lie on 3/8 of the interdistance, reckoned from the 73-inch point. To get other points with the boundary value 70 one

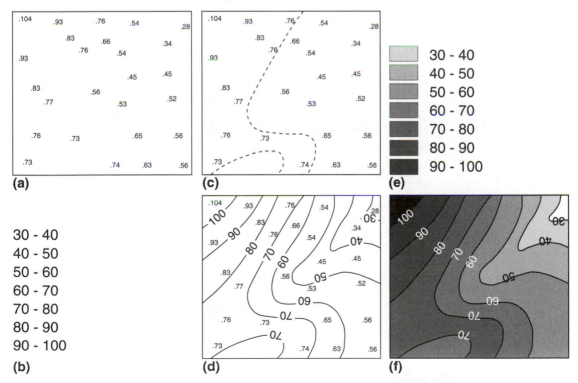

Figure 7.24 The production of point data-based isoline map (after Monkhouse and Wilkinson, 1971)

would have to proceed in a similar way between other data pairs on the opposite sides of 70. As a next step, shown in Figure 7.24(c), all constructed points with the 70-inch value are connected. In Figure 7.24(d) all other boundary lines have been constructed as well. Finally, in order to better perceive the general trends, in-between the isolines grey value tints are added (Figure 7.24e) in much the same way as in choropleth maps. Because the values increase continuously, more than seven classes would be acceptable in the final map (Figure 7.24f).

When the original data have not been collected or measured for point locations, as for weather stations or heights, but for areas, the first steps of the production procedure would be different. Data collected for regular grid cells (these can be population numbers), as in a grid map, are totalled per grid cell, and then assigned to the grid cell centre. These centres are then used as the data points, and from here the procedure is similar to the one described above (Figure 7.25).

There has been quite some opposition to this application of the isoline method, to the point where area-based isoline maps have been called pseudo-isoline maps in continental Europe (in the UK they are called isopleth (*iso* for 'equal' and *plethos* for

'value'). The point made against them was that discrete data (data valid for specific enumeration areas) were treated as if they were continuous. But this is all a matter of definition. If population density is not regarded as the ratio between the number of people living in a specific area and the size of that area, but as the number of people within a standard area size, such as a circle with a surface of 1 km^2, which can float over the area, then the concept could refer to a continuously changing value. This floating circle (to be perceived as a circle drawn on transparent material with which a population dot map is scanned) method was used to produce map Figure 7.22(c).

The prime issue regarding data-point-based isolines is the representativeness of these point locations regarding the phenomenon they describe for their surrounding area. The more homogeneous the surrounding area, the more representative the data-points.

In contrast to choropleth maps, isoline maps show us trends; they show clearly in which direction values for the phenomenon being represented are increasing or decreasing. Because of this characteristic, they are very well suited for comparing different phenomena and assessing whether there are correlations between

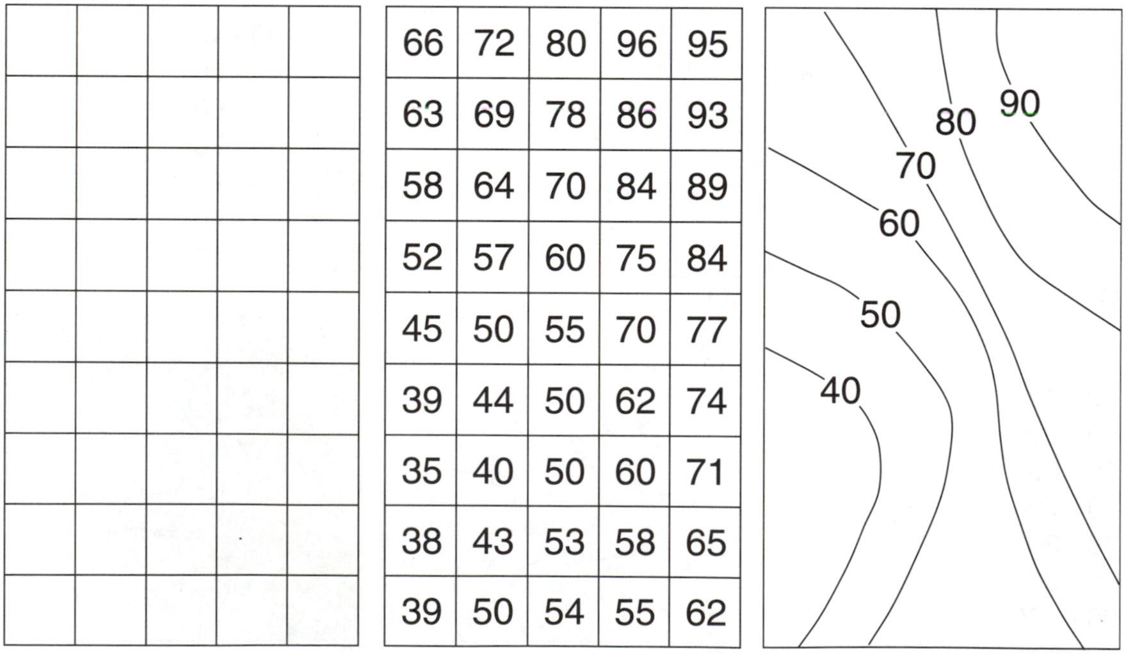

Figure 7.25 The production of isopleth maps through interpolation between grid cell centre values

these phenomena. In this regard they perform better than choropleth maps.

7.5.4   Nominal point data

Nominal data valid for point locations are represented by symbols that are different in shape, orientation or colour. There is a general division between figurative and geometrical symbols, the figurative ones being used when associations might ease recognition of the symbols. For more abstract phenomena geometrically shaped symbols are used. Figure 7.26 shows some associative figurative symbols. The more elaborate their shapes, the more they will be subject to a tangle on the map, resulting in severe legibility problems. These figurative symbols will probably be the first map symbols map readers will be confronted with during their education, as figurative symbols (like grain sheaves for agriculture, wheels for manufacturing or cows' heads for animal husbandry) usually dominate cartographic material for primary schools and tourists (Figure 7.26). With geometrical symbols, better map legibility is coupled with less recognizability.

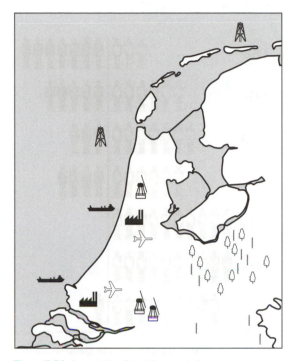

Figure 7.26 Associative figurative symbols

7.5.5   Absolute proportional method

Discrete absolute values, valid for point locations or for areas, can be represented by proportional symbols. For this purpose figurative symbols are not well suited, as their shapes are so complex that it is very difficult to compare their dimensions. Geometrically shaped proportional symbols are much better suited here. Figure 7.27 shows some examples of geometrical proportional symbol maps.

This is an appropriate place to mention that different graphical languages each have their own grammar. Next to cartography another graphical language is that of isotypes (using associative symbols, each with the same standard value), developed in the 1930s by Otto Neurath. The principal aspect of the grammar for his isotypes was that differences in value would be represented by differences in the number of symbols. Figure 7.28 shows an isotype of the number of Europeans with or without a constitution. In 1793, 30% had a constitution, a third of which were within a monarchy. Another aspect of the grammar is the central vertical line, which indicates the shift from unconstitutional to constitutional rule.

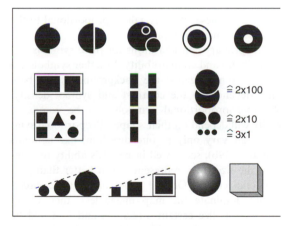

Figure 7.27 Geometrical proportional symbols

The same principle of showing differences in value by different numbers of symbols cannot be used on maps. Figures 7.29(a) and (c) show this: the large number of symbols obliterate too much of the map, and thereby threaten to block the geographical background, the very reason for showing the data on the map. Therefore, different values in cartography are represented by symbols differing in size. The areas

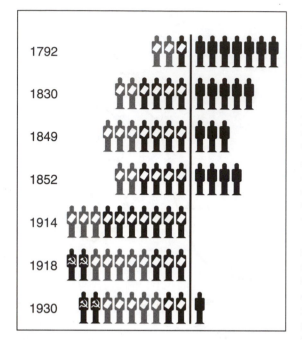

Figure 7.28 Isotype of the proportion of Europeans with a constitution (from Neurath, 1930)

covered by these symbols will be proportional to the values they have to represent.

The primary considerations for these symbols will be legibility and comparability. Whether symbols are legible or not against the background of the base map depends on the contrast and symbol density. Whether proportional symbols can be compared easily will depend on their shapes. Proportional symbols that vary only in one direction, like columns (Figure 7.29b), score well in people's ability to compare the sizes they represent, much better than proportional circles. On the other hand, these circles would dominate the map image less, they would not monopolize certain directions and it would be easier to apply them within the areas they relate to. As many map-readers underestimate continually proportional circles when comparing them, graded circles can be used instead, which only have a limited number of symbols to denote specific size categories.

The range concept, introduced in Section 7.4, is very important in the context of proportional symbol maps, as it refers to the ratio of the highest and lowest value that can still be represented proportionally, without impairing legibility. As could be seen from Figure 7.12, the range between the smallest proportional circle symbol that can still be perceived and the largest one that can be applied to the map without disturbing the map image too much is 1 : 2500. As the dimensions are proportional to the square roots of these values, this denotes the difference between a 1 mm diameter circle and a 5 cm diameter circle. When one tries to visualize a similar range through proportional columns, and the smallest value is rendered through a 1 mm high column, the highest value would have to be rendered through a 2.5 m high column, which obviously would be impossible. The only way to render such extremes would be by introducing a threshold value below which all values would be represented by an asterisk, for example, thus reducing the range.

It has been proposed to use three-dimensional symbols in order to increase the range that could be represented on a map. The idea was that by constructing three-dimensional symbols, the dimensions of these symbols would be proportional to the cubic roots of the values, instead of to their square roots as in two-dimensional symbols. When compared to proportional squares or circles, the space taken by 3D symbols will thus be reduced considerably. Perception research has shown, however, that with 3D symbols simulated on paper maps, the surface the model actually covers on the map is taken into account instead of the 3D model, thus leading to serious underestimations of the actual values represented. Thus the use of 3D symbols is only advocated here in situations in which a better 3D image can be generated, either through anaglyphs or stereo images (Kraak, 1988).

When used as symbols for areas, proportional symbols tend to be perceived as ratios: it is the ratio between the part of the area covered by the proportional symbol and the area itself which is taken account of. This phenomenon is most clear when proportional symbols are rendered in grid cells. Here the impression generated by them is similar to that of grid-cell choropleths (Figure 7.22b and e).

The production procedure to follow in this special case (of using proportional symbols for designating area characteristics in a regular grid) can be explained on the basis of Figure 7.30. In this figure (housing density in Europe) absolute area values are related to the number of regular grid nodes that fall into the respective areas: if 300 000 houses (eastern France) belong to an enumeration area in which 12 nodes fall, each node will have to represent 25 000 houses, and will be proportioned accordingly. As

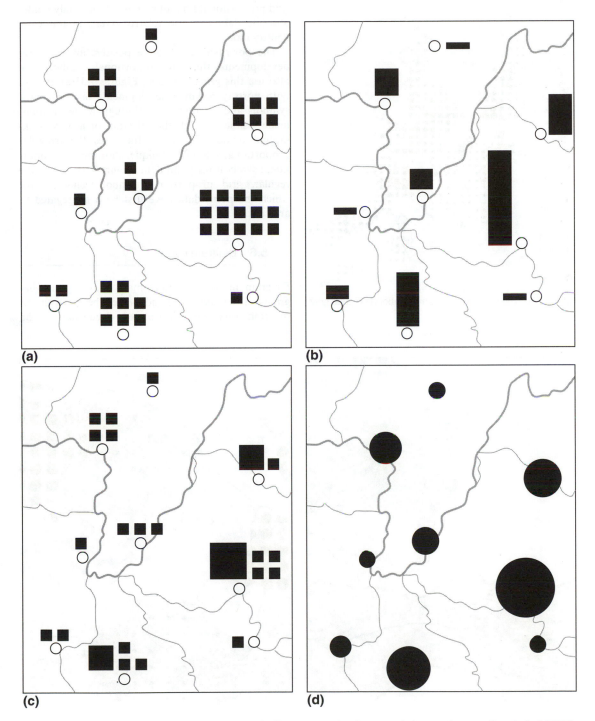

Figure 7.29 Comparison of the disturbing influence of different proportional map symbols upon the map (from Imhof, 1972)

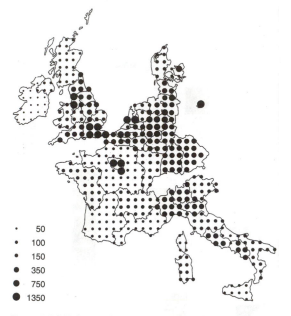

Figure 7.30 The use of proportional symbols in regular grids

each node also relates to an area of a specific size, this grid proportional symbol map would not only render absolute data but also relative data – in this case, density data.

The portrayal of absolute positive and negative developments through proportional symbols can also use this grid technique. Figures 7.31(a) and (b) both portray negative developments through black symbols and positive ones through white symbols, made visible through the selection of a grey background colour. Figure 7.31(a) uses the irregular proportional symbol technique, showing the proportional symbols for point data. Figure 7.31(b) uses the regular, grid proportional symbol technique, for which the point data have first been aggregated to area data.

7.5.6   Diagram maps

Diagram maps are simply maps that contain diagrams. Their use is not advocated here. Diagrams function very well in isolated circumstances, on

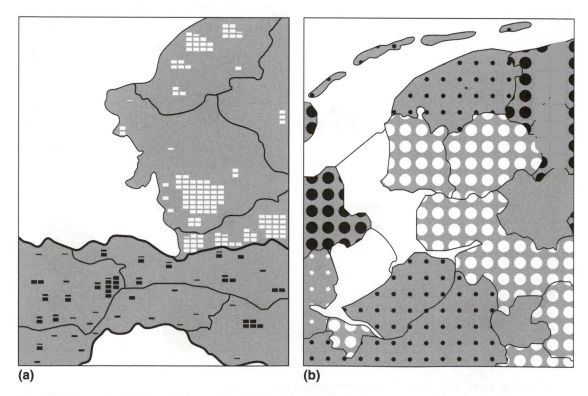

Figure 7.31 Proportional point symbols used for point data, showing absolute increase or decrease in a specific time period

their own or in pairs, in allowing comparisons between figures, or in visualizing temporal trends. But when represented against the background of a map there are usually too many distracting circumstances: the map background, geographical names, the fact that they are not situated on the same line any more, etc.

Nevertheless, diagrams are much applied to maps. One discerns between *line diagrams* in which the temporal trend in a phenomenon is indicated (such as yearly temperature of precipitation averages); *bar graphs*, in which the length of the column has been subdivided proportionally between various characteristics; *histograms* (in which a number of contiguous columns are used); *area symbol diagrams* (such as pie graphs) in which the area of the diagrams has been subdivided; *flow diagrams* (which will be covered in the next section); and *area diagrams*, in which the whole map area has been subdivided according to the percentages of the various characteristics discerned.

An example of the area diagram map is given in Figure 7.32(b). The basis shown in Figure 7.32(a) is a dot map, indicating the location of two language groups, A and B. In the area shown, there are 29 000 A and 56 000 B, so in this area the A group forms $(29\ 000/(29\ 000 + 56\ 000)) \times 100 = 34\%$ of

the population. This then should be shown in the area diagram map. If the area were subdivided into 100 grid cells, 34 of them would be designated A and 66 of them would be designated B. Though this would enormously distort the actual geographical image, by rendering the correct proportions, the geography might be helped a little by locating the designated grid cells in such a way that at least the actual distribution patterns are simulated somewhat.

The phenomenon of diagram maps is not to be confused with maps that have diagrams added to them in order to substantiate the categorization that has been applied. In a choropleth map the classification can be sustained by showing in a histogram the number of observations that fall within each class. In a scatter diagram or triangular graph, a categorization on the basis of three characteristics that together make up for the whole data set can be sustained.

The negative advice at the start of this section is based on the consideration that a map is not the proper environment for diagrams because they contain too many distracting graphical cues. The only advantage of adding diagrams to maps would be to put them into their geographical context. However, with proper reference to their actual location, a series of diagrams would be much better placed next to

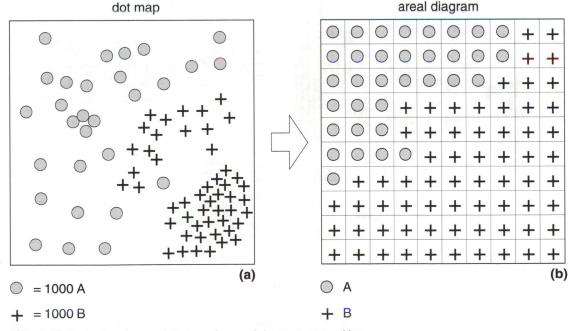

Figure 7.32 Production of an areal diagram of an area's language composition

each other outside the map. Here, because of their adjacency, they can be better interpreted. In a multimedia package they could be consulted outside the map environment.

When cartographers or researchers produce diagram maps, it is mainly for analytical purposes. The data are being analysed in their proper position, which will help the analysis – but as an inventory map it would not be suitable for communication. A good example is the composite migration diagram shown in Figure 7.33. Here, the positive and negative net migration is visualized for 10-year periods for cantons in western Belgium. For each individual canton the diagram shows admirably the trend in the population development, but together the diagrams are too complex to show a clear regional trend, which is why the map has been analysed. As a result of this analysis a number of migration types has been discerned. There are cantons with a continuous net immigration and with a continuous net emigration, there are cantons that lost many inhabitants in the second decade of this century (why?), there are cantons that lost people early in this century and gained later. Nearly all cantons can be assigned to one of these types, and visualized accordingly. The resulting image shows much better the actual distribution of

the various types, and therefore of the migration trends.

7.5.7 Dot maps

Dot maps are a special case of proportional symbol maps, as they represent point data through symbols that each denote the same quantity, and that have been located as well as possible in the locations where the phenomenon occurs. In the case of a population dot map where each dot represents one person, it would be possible to produce a dot map with absolute correct locations of the dots. Whenever people have been aggregated because the representative value of the dot is not 1 person but 5 or 100, it becomes impossible to show the locations of the persons represented correctly and approximations have to be made. A solution would be to put the dot in the centre of the addresses of the inhabitants it represents, or in their gravity point. The dot locations have to be selected in such a way that they characterize the actual population on distribution.

So dot maps show patterns: for instance, concentration and dispersion of the population distribution,

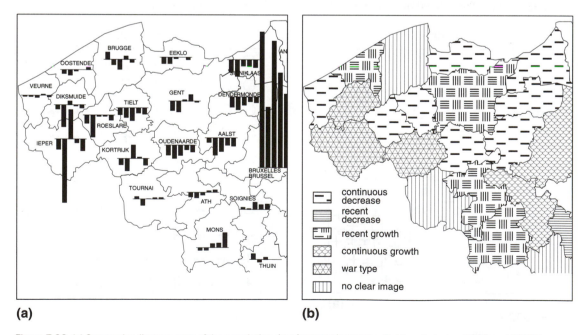

(a) **(b)**

Figure 7.33 (a) Composite diagram map of the population development in western Belgium between 1900 and 1950 (source: *Atlas van de Nationale Survey*), for analytical purposes, and (b) chorochromatic representation of the same information for communication purposes

in a population dot map, or subgroups of the population, as the students represented in Figure 7.22(a).

A population dot map is usually generated according to the following procedure (Figure 7.34): population data will be available for enumeration areas (Figure 7.34a), and generally additional information will be available in the form of topographic maps, land use maps or remote sensing imagery. On the basis of this additional information (Figure 7.34b) it will be possible to determine which areas are uninhabited, which are sparsely inhabited and which have a high population density (as the contiguously built-up urban areas). So on the basis of the data and the additional information, the enumeration areas will be broken down into smaller units considered homogeneous as regards their population distribution, with the population numbers that are supposed to apply

to them (Figure 7.34c). The final step will be the translation of the values to be represented into graphical form, by choosing a dot with a specific size and a specific representative value (Figure 7.34d), and by applying the dots regularly over the areas considered homogeneous.

Without the additional information one would only be justified in applying the number of dots calculated for each area regularly over that area; with the additional information one can be more specific.

When the size of the dots is too large, a situation will quickly occur in which the dots have a tendency to coalesce or merge, so that individual dots can no longer be distinguished. Though this is not serious for a restricted number of locations (after all, it is never the intention with dot maps to show densities

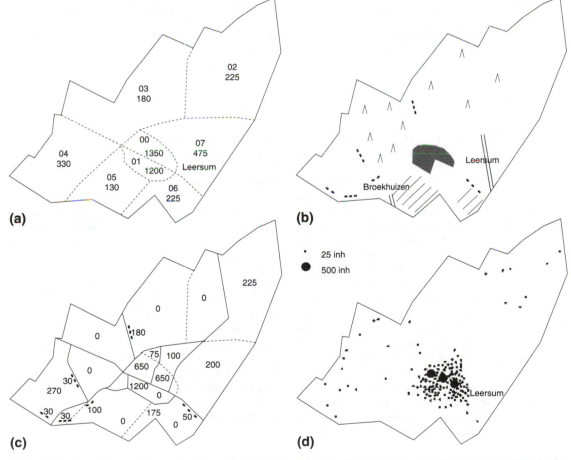

Figure 7.34 Production procedure of a dot map: (a) administrative units; (b) topographic map; (c) combining administrative units with the topography; (d) the dot map

or to have the dots counted), it is better to avoid this if they would coalesce over larger areas.

When the representative value of the dots is too large, fewer dots will have to be applied to the map, but in sparsely inhabited areas large tracts will go without any dots at all, so no pattern will show here. When the dots have a small representative value, many dots will have to be located, and there will again be the danger of merging. When the differences in density are just too large for rendering both rural and urban densities, one may be compelled to use an additional dot size for larger urban values (Figure 7.34d).

Because of these considerations the actual dot size and representative values have to be tested out for a few different areas in order to check whether the required impression will result.

A number of software packages are equipped with the facility to produce dot maps. But as these work on the basis of random generators of the dot locations within the areas specific, there is not much use in applying them, as the resulting image will bear little resemblance to the actual distribution within the areas.

7.5.8   Flowline maps

This is one of the few map types that simulates movement. Movement can be simulated on static paper maps in a number of ways: by using graphical variables that give the reader an ordered impression (through differences in size or in grey value); by showing a number of situations adjacent in time next to each other (the filmic method, as in a comic strip); or by using symbols that are associated with movement. Flowline maps use the third way, as they use arrow symbols. This is the most useful symbol in cartography, as arrows indicate both the route along which a movement occurs and the direction along the route (by the way in which the arrowhead points); also the volume transported along that route can be shown by the relative thickness of the arrow's shaft. So far this definition is valid for flowline maps, flowline diagrams and for vector maps (see also Figure 7.15). *Vector maps* only show the size of the forces that occur at specific points in specific directions. Wind velocity maps are an example. *Flowline maps* show the specific route of the movement or transport as well, but are not further subdivided. *Flowline diagrams* are further subdivided; for instance, in the amount of goods transported from A to B, a flowline

diagram may show the proportion of those goods that have been transported by boat, by dividing up the arrow shaft longitudinally into different sections denoting different transport modes.

The impression given by proportional arrow symbols is one that is governed by both the length of the route and the thickness of the arrow, i.e. the amount transported. This product of thickness and length really is a transportation achievement impression rather than an impression of the amount of goods transported (Figure 7.35a). If it is only the amount of goods transported from all over the world to a given location that has to be visualized, this can be done better by expressing these amounts proportionally at their places of origin (and eventually linking these symbols to the destination by arrows showing the direction, as in Figure 7.35b). The amounts can be compared in this way more easily than from arrow symbols pointing in all directions.

Whenever traffic in both directions along a route is in balance, there is not much point in adding arrowheads, and thereby the symbol changes into a band-like symbol, still proportional to the amount of goods or persons transported along it.

7.5.9   Statistical surfaces

The three-dimensional representation of quantitative data such as used in choropleth and isoline maps for analytical purposes can help in their two-dimensional representation. Such a three-dimensional representation may be called a data model or a statistical surface. In the case of the theme being mapped as height over sea level, the data model will simulate tangible reality (Figure 7.6(b), 7.10 and 7.16(e)). Where it portrays other phenomena the data model will serve as a yardstick for assessing whether, through classification procedures, a correct view of the data has been generated. Of course, the data model can be used on its own for data communication, but it has some disadvantages here: it is not generalized through classification and therefore might present too complex an image, and because of its relief, some areas might be obliterated by peaks in the three-dimensional data model. Because of its perspective, it would be impossible to read exact values from the map. There would be advantages as well: the image generated by a three-dimensional data model is a very dramatic one, which will be remembered for a long time, and it would present a good overview of the general trend of the data.

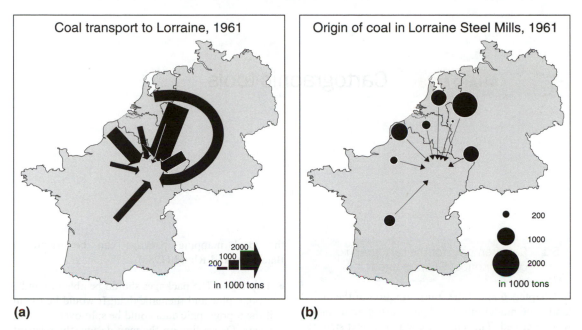

Figure 7.35 Flowline maps showing transportation achievements; maps for comparison of the actual coal quantities transported from various destinations to Lorraine

CHAPTER 8

Cartographic tools

8.1 Requirements for the cartographic component of GIS packages

In Chapters 6 and 7 the transformation of the digital landscape model into the digital cartographic model was described. This chapter covers the next step: the transformation of the digital cartographic model into the map, be it permanent or virtual (Figure 8.1). This used to be a time-consuming aspect of cartography, but has now been reduced to mere button-pushing when output is required, from which the printing plates will be made. So with the exception of a diazo-machine only the digital processes will be covered here, as only these are deemed relevant for GIS-users.

The functionality of software available for the production of maps varies strongly. The larger GIS packages have very extensive cartographic modules, with options for map design and production, while others have limited capacity to visualize GIS data. Smaller packages are often more limited. Each year, the *GIS Sourcebook* (1995) publishes an overview of the capabilities of most packages available. GIS Europe and GIS World regularly publish overviews of software functionality and peripherals. However useful as a starting point, these overviews only indicate what is possible, and do not tell the reader how, and with what options, the software realizes its functionality.

If an organization has to select the (GIS) software to produce maps, some specific aspects regarding cartographic functionality have to be considered. These reflect general use, design and output. It should be realized that the organizational requirements will determine which of the points below should be emphasized most. A more extensive discussion on the ideal mapping package can be found in Blackford and Rhind (1988).

- *Map types.* The packages should be able to handle topographic and thematic data. It would be useful if the topographic data could be split over different layers. Depending on the map design, the relevant layers can be switched on or off. Examples are layers with roads, boundaries and administrative units. Thematic map types to deal with qualitative and quantitative data should be available. Examples are choropleth and chorochromatic maps. A link with a database is preferred, because it allows one to use the same map design for different (temporal) data without too much effort (see Chapters 5 and 7).

- *Marginal information.* The software should have (interactive) facilities to create a legend, north arrow, scale bar and map title. Often the legend is created by default, but users should have the possibility to choose their own symbology and colours. For a scale bar and a north arrow many packages have a symbol library available from which the users can use the most appropriate representations. Since many organizations would like to have their logo included on the map as well, the option to add user symbols to the library is appreciated. For titles and other texts many fonts are often available. Some packages do not display any marginal information at all when dealing with screen maps, because screens are relatively small. In these situations the user can sometimes access the marginal information through pop-up menus. In other words, the information is available on demand.

- *Data manipulation.* Options to change the coordinate system, to generalize topography or to classify

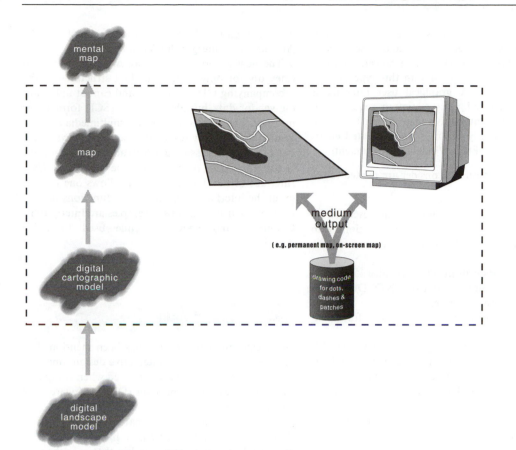

Figure 8.1 From digital cartographic model to permanent or virtual maps

attribute data are needed. For most GIS software this is no real problem, since these facilities are seen as generic GIS functions. However, generalization is often limited to a single line generalization algorithm, and the possibility to classify attribute data will depend on the database linked to the GIS package. In those packages devoted to mapping only, data manipulation facilities are often limited, and the user can, for instance, choose between three data classification methods only (see also Section 7.3).

- *Output.* Facilities to send maps to raster or vector plotters and to use different monitor screens should be available. It should be possible to visualize the maps designed on any output medium. Most screen types are made available via the software and operation system environment which have so-called drivers available for each screen type (an example is the Microsoft Windows environment). The same principle is valid for those

devices creating permanent maps, like printers, plotters and even film writers (see also Section 8.3). Since current printing presses can print only one colour at a time for each printing plate, in order to reproduce all possible colours one needs at least four printing plates (for the primary colours and black; see Section 6.3). The preparatory process that leads up to these four printing plates is called colour separation. When it comes to automatic colour separation only the larger GIS packages, and the generic desktop publishing packages, can handle this. Some of the larger GIS packages such as Intergraph's MGE have dedicated modules for cartographic production: Map Publisher and MapFinisher. For others, like Arc/Info, such modules are provided for by third parties.

- *Graphical user interface.* A big problem of most GIS-related software is their user interface. If users agree upon one issue, it is the lack of userfriendliness of GIS-related software. If one realizes the enormous number of commands and options available, this is not strange; however, the time when only experts worked with this type of software has passed. A graphical user interface is nowadays regarded as a minimum requirement, especially when interactive map design has to be executed. Currently the most sophisticated environment is the desktop Windows environment.

Some general questions to consider are related to the following issues:

- *Quality*. How is the functionality, as described by the vendor, realized? What are the relations with cartographic theory? What kind of output options are available?
- *Usefulness*. Does it fit into the organization's computer environment (PC, Mac or UNIX-DOS)? Is it compatible with other available (GIS) software packages (to allow for data exchange, for example).
- *Userfriendliness*. How easily will a new user be able to work with it? What level of training is required? What are the manuals like? How good is the support by the vendor? How are errors reported and processed?
- *Cost*. How much has to be spent (initial cost and maintenance)?

Tomorrow's users of geographical information systems will require a direct and interactive interface to their geographic data, which will allow them to search for spatial patterns, steered by their knowledge. In this process the on-screen map plays a key role as an interface to the data, and it becomes central to spatial data handling and analysis activities. The map itself will become a process. Today's representatives of this evolution are the so-called dataviewers. All of the major GIS software vendors have one or are developing one. Examples are ESRI's ArcView, Intergraph's Vistamap and Tydac's SpansMap. Dataviewers are map-based data browsers. The first generation of these products allows the user to look at data prepared and structured in the vendor's major GIS packages. The user is guided by a generic graphical user interface that allows the display of maps, and provides access to the data behind these maps. Often it is also possible to integrate raster imagery. Some basic cartographic functions are available to classify and symbolize the data. Although conversion to paper maps is possible, its main objective is to function as an on-screen dataviewer. Figure 8.2 shows an example of ESRI's ArcView and Intergraph's Vistamap.

The new generation of dataviewers is more open (they do not only match the data structures of the accompanying GIS packages), and cannot only handle vendor data but also data in ASCII format, for instance. They also open toward multimedia and the general desktop environment. The map allows direct links to text processors and spreadsheets. Important is the possility to customize the Graphical User Interface of the dataviewer. This allows one to program the interface with only those functions needed for particular user groups. Examples are Intergraph's Geomedia and ESRI's Avenue used to tailor ArcView 2.0

8.2 Desktop mapping

On-screen interactive design has been mentioned in the previous section. The interactive design functionality of most GIS software is limited when compared with some desktop publishing (DTP) software. For instance, positioning text in a GIS map is done through input of its coordinates in a macro-file. When shown on the screen, it might have to be moved a little up or down, and in that case the user has to edit and run the macro again. Graphics produced with software such as CorelDraw or Freehand would allow the user to drag the text to the correct position with the mouse, and even scale it interactively. Also, the application of specific design and construction functions available in DTP software gives the graphics a fresh and sophisticated look that is missing or very difficult to create in most GIS maps, even though these GISs do have extensive cartographic functionality. Figure 8.3 shows the difference between a default GIS map and the sophisticated version produced from it, enhanced by DTP packages. DTP software, on the other hand, is only about graphics, and elementary options such as input of geographic boundaries is often problematical. Most packages offer libraries with maps, just as they offer libraries with flags, animals, cars, airplanes and other clipart symbols, and have many text fonts available as well (see Figure 8.4 for some examples). Even CD-ROMs with cartographic clipart in different graphic formats are offered on the market. Such maps are good if one needs just the outline of a

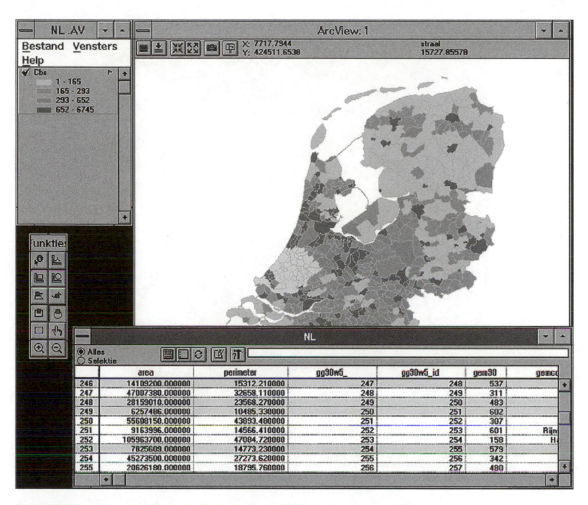

Figure 8.2 Dataviewers: (a) exploring the Netherlands with ESRI's ArcView (Courtesy ESRI)

country, but are often of little use if one needs a proper base map.

Currently, exchange of graphic files is often limited to bitmaps, and this traffic would be one-way only, which restricts the number of graphic operations that can be executed. For those DTP packages that can import Postscript (a page-description language that functions as an *ad hoc* graphics standard as well), a vector exchange would be possible as well. In that DTP environment the user can decide on the final layout of the map. From the same environment the output media can be reached directly. Currently a return to GIS from DTP is limited to bitmaps. It is, for instance, not possible to produce a map with GIS software, transfer it to a

DTP environment, enhance its graphics, return it to the GIS environment and re-establish the direct link with the database.

Problems one can encounter in the production of a thematic map are described by taking an example where the objective is to produce a map of the United States' population density, on a county level. The design has to adhere to a specific layout including an organization's house style. What options would be available? GIS packages as well as specific cartographic packages can produce such a map from their database, but not in the layout required. How can DTP packages help? Let us first consider the topographic base map. The easiest solution would be to export the base map from the GIS

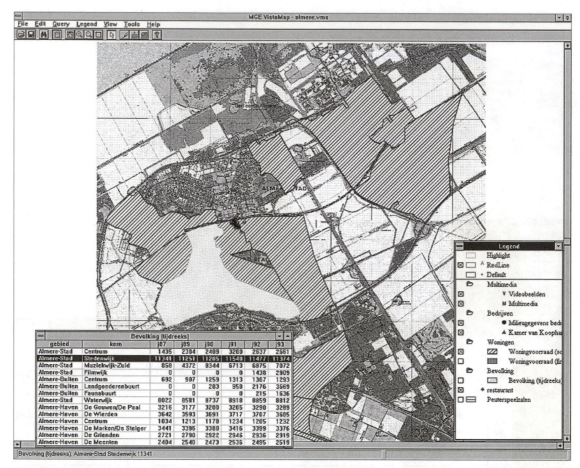

Figure 8.2 Dataviewers: (b) visiting Almere (the Netherlands) with Intergraph's Vistamap (Courtesy Intergraph)

packages into the DTP package. This would result in a boundary file ready for use. However, when this option is not available, a time-consuming process starts, especially when only bitmaps are available. A bitmap could be the result of scanning an appropriate paper map. In this situation one has to apply heads-on digitizing to create a vector map with over a thousand polygons representing the individual US counties. Some DTP packages have tracing options available, but often these do not result in the required geometric accuracy. A specific problem is caused by internal boundaries. These lines belong to two individual counties (polygons), and have to be digitized twice, but the lines have to match exactly otherwise the result will not be acceptable. Some DTP map libraries provide vector line maps. From these, polygons have to be created in order to be able to assign each county the specific colour or pattern of the class

it belongs to. When the DTP libraries contain polygons, the maps are ready for use.

Adding thematic content to the map can be effectuated together with the conversion of the base map from the GIS map, via for instance Postscript. In all other options described above it will be necessary to click each of more than a thousand polygons to assign them their colour or shading pattern. The desktop functionality can also be used to adapt the map layout to the layout typical of the organization (house style). Depending on the design, this again could be time-consuming. What if the final map has to be updated because new census data have become available? With most software this means clicking all counties that have changes again, because importing the map from the GIS means going through the special design efforts again. Some of the DTP packages have the facility to link with spreadsheets. From

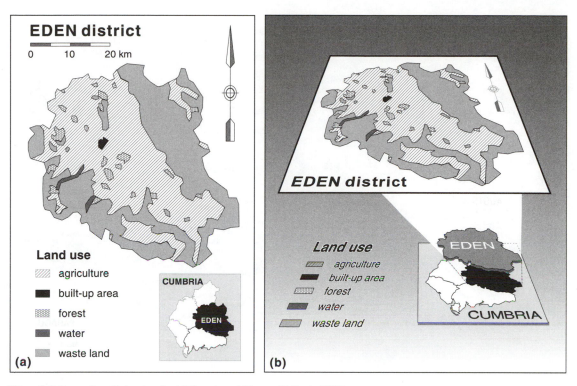

Figure 8.3 Examples of (a) a standard GIS map and (b) a sophisticated DTP map

these spreadsheets the maps can be updated without much effort. To export data from the GIS into a spreadsheet is no real problem. However, a link between polygons and a location in the spreadsheet has to be established. With an appropriate macro language this problem can be solved.

How can we enjoy the best of both worlds? A trend gaining momentum is the ability within the environment of a specific graphical user interface, such as Microsoft-Windows, to paste and copy graphics from one application into another. In the near future the desktop environment will be enhanced even further. When dealing with a document including text, graphs, tables and maps, clicking on a table will make spreadsheet functionality available, clicking a map will activate GIS or graphics software, and all changes can be saved in the original document. This will be the strength of the new desktop environment in which geographic data can be as easily dealt with as today's texts and tables.

Software which comes close to the qualifications described above is the Cart/o/Info development environment. Based on the Cart/o/Graphix desktop mapping software, it is a toolbox for building cartographic information systems. Systems constructed with this toolbox offer the user fast access to georeferenced raster maps, vector maps, many thematic mapping options and some GIS analysis functionality, all data-driven from relational databases and/or spreadsheets.

8.3   Map production

This section is organized according to the scheme in Figure 8.5. It splits the map production process into five distinctive phases. The text will concentrate on the digital approach, but will provide parallels from conventional procedures as well, because in many environments one is in the transition between analogue and digital map production. The scheme does not pretend to be the only possible or complete approach, but presents a valid generic method of map production. Many lines appear between the boxes representing both the digital and conventional

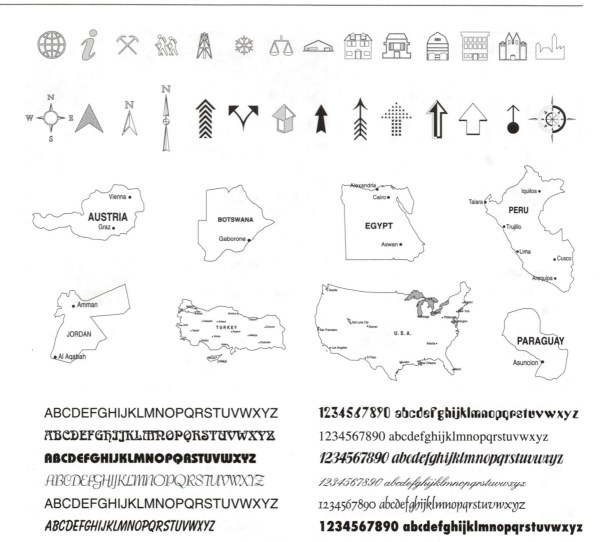

Figure 8.4 Desk top publishing packages: available fonts and clipart for cartographic purposes

approach. They indicate the possibility of switching from the analogue to digital methods and vice versa. Again, such lines are only an indication, because technology evolves so fast that a new line could be added on a weekly base. Other cartographic textbooks, such as Keates (1989) and Robinson *et al.* (1995), and ICA's Basic Cartography series (Anson and Ormeling, 1993–1996) go into much more detail, especially regarding the conventional approach. The phases distinguished in the scheme are preparation of the design, the map specifications, creating the map image, the reproduction, and the final product.

The first phase is the preparation of the map design. Data will be selected from one or more

(edited) Digital Landscape Models. In a GIS environment this phase will be steered by the objectives of spatial analysis operations. The selection will include topographic base data and thematic content representing the results of spatial analysis. In a conventional environment a sketch map would be produced.

Defining the map design specifications is the second step. It includes the definition of the Digital Cartographic Model, which will contain all graphical attributes (such as line colour, shading pattern and text fonts) of the geographic objects to be represented in the map. A similar process takes place in the conventional environment where the sketch map is converted to a map that includes the information similar

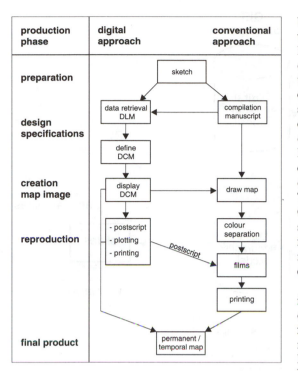

production phase	digital approach	conventional approach
preparation	sketch	
design specifications	data retrieval DLM ← compilation manuscript / define DCM	
creation map image	display DCM	draw map
reproduction	- postscript - plotting - printing	colour separation
	Postscript → films	
	printing	
final product	permanent / temporal map	

Figure 8.5 Map production scheme: five production phases can be distinguished, and for each of these digital and analogue options are given. It is possible to switch from digital to analogue and vice versa at almost any phase

but more importantly, it is possible to keep a link with the GIS database. The map can be queried. More in-depth or additional information can be obtained by the viewer from the map. For instance, the real value of an individual geographic unit in a classified choropleth can be accessed. It is possible to select a group of units and ask for statistics. Maps created in a data analysis or data exploration environment are just intermediate maps for individual use. Here less or no attention is paid to the design. In a conventional environment one will draw the map, often on paper, a transparency or engraving film. Tools used are pens, rulers and engraving equipment. Only if one intends to limit the production to a single simple hand-drawn map will this phase result in a final product; otherwise the third phase is the beginning of the map reproduction in a conventional environment.

Phase four prepares for multiplication. If one intends to keep the map image in digital format only, one might store it as a computer graphics metafile (CGM, a file that contains the instructions to reproduce a vector image) or in any other format. It can be made available on networks like the World Wide Web. Other options would be to incorporate the map in an electronic atlas or in an animation or multimedia environment. If the digitally produced map is to be reproduced on paper, conversion software and hardware have to be used that transform the DCM into a tangible set of final combination films (one for every printing colour) or even transfer the map image directly onto printing plates.

From Figure 8.6 it can be seen that many different techniques are available for displays on paper. The choice will depend on the type of output needed. The size of the paper, the use of colour and the number of copies required will help to decide what type of equipment is needed. It should be realized that a pragmatic factor such as the price of the equipment is influential as well. It may vary from 1000 ECU for a simple black and white A4 laser printer to more than 50 000 ECU for a colour A0 electrostatic plotter. Also, one should not only consider the initial cost, but the price per copy as well.

Software used to design maps will normally translate its drawing commands into a special language. The most common of these languages is Postscript, developed by Adobe. The Postscript code contains a resolution-independent textual description (normally in ASCII) of the map. A polygon, for instance, is described by a list of coordinates and some codes that can provide information on the polygon's inter-

to a DCM. The layout of the maps represented by DCMs is inspired by the contents of Chapters 6 and 7.

The third step consists of the display of the DCM. The map is shown on a screen. This medium has its limitations in size, resolution or colour. The map image displayed can be used to check if it looks according to the specifications and/or expectations. Phases one, two and three are part of an interactive design process. During this process one deals with questions such as: do we have to include additional data or should we delete some to keep the map legible and informative? Do the colours of the symbols match harmoniously? Do they contrast sufficiently? Is the line thickness correct? Are the place-names positioned correctly? If changes have to be made one can return to phases one or two as the lines in the scheme show.

The on-screen map could even be the final product. Maps on screen in a GIS environment can have additional capabilities that are of great value. Not only could their production be effectuated interactively,

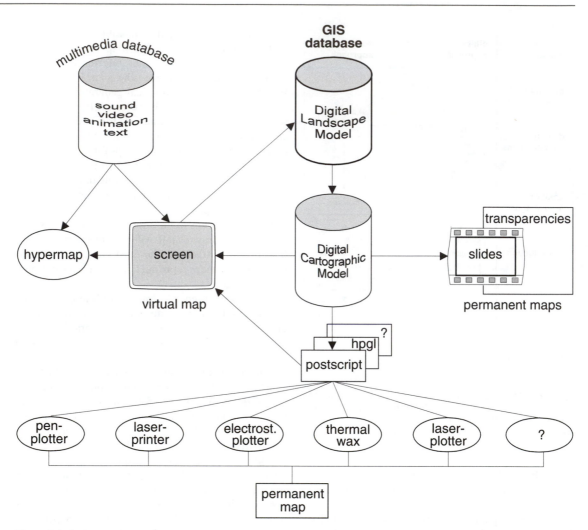

Figure 8.6 Options to display permanent and virtual maps. Based on the digital landscape model (DLM), a digital cartographic model is prepared (DCM). From the DCM soft copies (the virtual map), and hard copies (the permanent map) can be made

ior, line type and thickness. Theoretically it should be possible, when using Postscript, to send the same map to devices with different output qualities. Those devices capable of processing Postscript will transform the Postscript code via a raster image processor to a bitmap, which can be put on paper. The Postscript code can be viewed on screen before it is sent to paper to check the result. Among the (raster) output devices that can handle Postscript are laser printers, electrostatic plotters, thermal wax transfer printers, ink-jet printers, and laser plotters. Laser printers work by an electro-photographic process, mostly in black-and-white. Their resolution varies from 300 dots-per-inch to 1200 dpi, on paper up to

A3 size. Electrostatic plotters are very fast machines that can handle paper up to A0 size, and can handle full colour. Thermal wax transfer printers use heat to transfer (colour) ink to a special type of paper and can handle up to A3 size. Ink-jet printers can also handle colour and different sizes of paper. Laser plotters are often used to produce films to be used for printing plates. It could be said they are the reverse of a scanner, and can handle resolutions from 1200 to 4800 dpi.

It is not only Postscript that is used to activate printing and plotting devices. Another *ad hoc* standard is HPGL, the Hewlett Packard Graphic Language. Most laser printers can handle this lan-

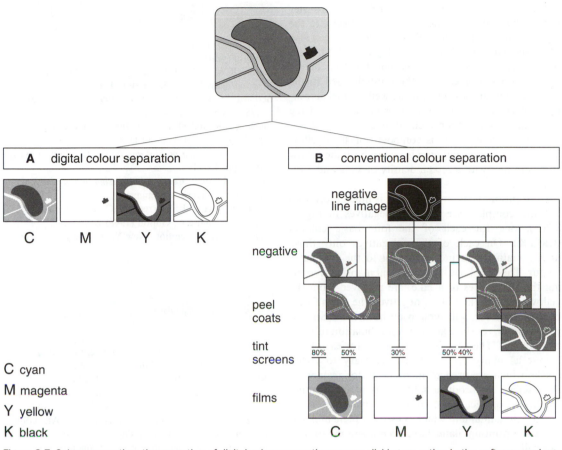

Figure 8.7 Colour separation: the execution of digital colour separation means clicking an option in the software package resulting in four files from which printing plates can be made. Analogue colour separation involves quite a complex process of splitting and combining several versions of the same or adapted image

guage but, more importantly, all pen plotters are activated by it. Since pen plotters are vector-oriented, HPGL in its most simple form contains pen-down and pen-up commands. Pen plotters do exist in types that can handle A4 to A0 sizes. As well as *ad hoc* standards such as Postscript and HPGL, many software packages have their own methods to activate output devices.

With current technology most of the output devices mentioned in Figure 8.6 are less suitable to produce many copies of the same map at reasonable prices. For this type of duplication, for instance to print topographic maps, one still relies on a printing press. Here the digital method currently goes up to plate-making, although experimental equipment for digital printing is available. Since a printing press can only print one colour at a time, colour separation is necessary (Figure 8.7). Normally one separates the colours of a map into four colours. A combination of the basic colours, cyan, magenta and yellow, (with black added for perfection (CMYK), allow one to print any colour.

If output has been plotted on transparent film, another reproduction method is feasible, especially for large sizes, i.e. diazo printing. Here light-sensitive dyes are used as chemicals that define the image. In a diazo-machine, light-sensitive paper is brought into contact with the transparent film during exposure. The exposed diazo compounds turn into nitrogen and a colourless rest product. The diazo compounds behind the lines on the transparent film remain intact. After development of these compounds (by ammonia vapours) a black, brown or grey image will remain.

Output devices such as the thermal wax transfer printer have internal mechanisms to effectuate the digital colour separation. Some DTP packages also have this capacity, as shown in Figure 8.7(A). Here the software creates four files, one for each of the basic colours. For each of the symbols used in the map the colours are known, as well as the percentages of cyan, magenta, yellow or black used. If the forest in the map is represented by a green polygon, and the shade of green is composed of 40% yellow and 30% cyan, this polygon is sent to both the yellow and cyan files with a screen density of 40% and 30% respectively.

Conventional colour separation is much more visible and complex, see the four negatives in Figure 8.7(B), one for each of the four basic colours (CMYK). On each of these negatives only those lines remain which indeed belong to the particular colour. The blue lines in the original maps, representing the shore lines of a lake, are removed from the yellow, magenta and black negative. The green forest border, however, will remain on both the cyan and yellow negatives since green is a combination of these two colours. The next step is to create for each basic colour as many peel coats as necessary to create masks which are used to incorporate for instance a 30% screen (to produce lighter tints of the printing colours). For each basic colour the masks are combined onto a final film which is used for plate making. After the printing plates have been made the presses can roll.

Further reading

Anon. (1995) *GIS sourcebook 1995*. Fort Collins: GIS World (yearly issue).

Anson, R.W. and F.J. Ormeling (1993–1996) *Basic cartography*, Vols 1–3. London: Elsevier.

Bernard, M. and P. Miellet (eds) (1993) Desktop mapping. *GIS Europe*, **2**(10), 24–46.

Blackford, R. and D. Rhind (1988) The ideal mapping package. In D. Rhind and D.R.F. Taylor (eds) *Cartography, past present and future*. London: Elsevier/ICA, pp. 157–168.

Keates, J.S. (1989) *Cartographic design and production*, 2nd edition. Harlow: Longman.

Robinson, A.H., J.L. Morrison, P.C. Muhrcke, A.J. Kimerling and S.C. Guptill (1995) *Elements of cartography*, 6th edition. New York: Wiley.

Products mentioned

Cart/o/Info (Cart/o/Info, Karlsruhe, Germany)
Cart/o/Graphix (Cart/o/Info, Karlsruhe, Germany)
ESRI Avenue (ESRI, Redlands, CA, USA)
ESRI ArcView (ESRI, Redland, CA, USA)
Intergraph Vistamap (Intergraph, Huntsville, AL, USA)
Intergraph MGE (Intergraph, Huntsville, AL, USA)
Intergraph Geomedia (Intergraph, Huntsville, AL, USA)
Postscript, Adobe (Mountain View, CA, USA)
SpansMap Tydac (Nepean, Ontario, Canada)

CHAPTER 9

Advanced mapping environments

9.1 Introduction

The present (1995) research frontier in cartography is in a number of advanced mapping environments where, because of the added potential of digital computers, new ways of visualizing and using spatial information are being developed. The new technological potential has led to conceptual and theoretical developments that are only now being addressed: if three-dimensional and animated presentation can be easily realized, then what is their effect on the viewer, what is the result of new graphical variables (Chapter 6) now available, such as shadow and volume in three-dimensional mapping, or duration and layered construction in animations, and what could be – in a multimedia environment – the added effect of sound? These are research issues cartographers are studying, and these go beyond GIS issues. In this chapter the effects and potential of electronic atlases, cartographic animation and multimedia will be covered.

9.2 Electronic atlases

Atlases are generally considered a higher form of cartography, as in their production there is both an extra planning and an extra structural dimension. It is not just one map that has to be ready at a specific time, but perhaps a hundred, and these maps all have to relate to one another. Atlases are intentional combinations of maps, structured in such a way that given objectives are reached. In a way, atlases are similar to rhetoric: if a number of arguments are presented in a speech in a given sequence, a specific conclusion is reached. Objectives of atlases may include the introduction of children to their environment (see also Plate 7), or to access global information in a reference atlas, but it may be just as well to provide awareness of specific environmental problems or to evaluate the availability of good factory sites.

Atlases work with a number of 'tools' in order to structure the information. An information hierarchy is arrived at by the use of a sequence tool and a scale tool: more important themes or areas are shown earlier in the atlas or at a larger scale than less important ones. By showing a number of thematic maps consecutively a causal relationship between these themes is suggested. Specific areas that are regarded most important can be highlighted by zooming-in on them in an inset-map. In the margins of the atlas maps there can be references to cities in other continents at the same geographical longitude or latitude, or to related themes depicted on other maps.

The analytical power of computers has given the atlas concept an extra dimension. A good example is the National Atlas Information System (NAIS) produced in Canada (Siekierska, 1993). In the maps in this atlas (which can also be accessed through the World Wide Web) all the items depicted can be queried: if the cursor touches an area, its name will pop up in a window; if a line element on the map, such as a river, is clicked on, its name and debit will be shown, as well as whether it is navigable or not. On the energy map in this NAIS, all power plants are displayed. Clicking on one of the point symbols that represents them will result in a display of all relevant data regarding this plant (type (nuclear/coal/hydro-electric/oil), capacity, name, etc.). So the maps in the electronic atlas function as an interface with the atlas database. This combination of database and graphical user interface (GUI) and other software functions developed to access the information is different from a GIS: special care is taken to relate all data sets to each other, to allow them to be experienced as related, to let them tell, in conjunction, a specific story or narrative. There will usually be a central theme (e.g. what has happened to the environment in the last 20 years, for instance; or whether all inhabitants of this country have equal access to its resources).

If traditional atlases are considered intentional combinations of maps, then not all electronic atlases might fit this definition, as there are several types of electronic atlases to be discerned:

- view-only electronic atlases
- interactive electronic atlases
- analytical electronic atlases

The latter two, discussed below, might be defined as intentional combinations of specially processed spatial data sets, together with the software to produce maps from them.

View-only electronic atlases can be considered as electronic versions of paper atlases, with no extra functionality, but with the possibility to access the maps at random, instead of the linear browsing that occurs in paper atlases. There is already a dis-tinct advantage over paper atlases and that is the cost of production and distribution. They are much cheaper to produce and – in form of diskette or CD-ROM – it is much easier to distribute (and update) them than paper atlases. Some extra aspects of easing their use might be the possibility (in a Windows environment) to view different maps together on the same monitor screen, by dividing it up. Examples of these view-only electronic atlases are the *Atlas of Arkansas* (Smith, 1989) and the *Atlas of Florida*.

Interactive electronic atlases are intended for a more computer-literate audience. They will allow their users to manipulate the data sets. The principle here is that there are no true maps: each map is a specific selection of data, processed in such a way as to come as near as possible to the essence of the theme's distribution, but it will always be biased by subjective elements. In an interactive environment users can change the colour scheme used for one of their own liking; they can adjust the classification method or extend the number of classes at will. Examples that allow this interactivity are the electronic atlases of France produced by RECLUS in Montpellier, France (Figure 9.1).

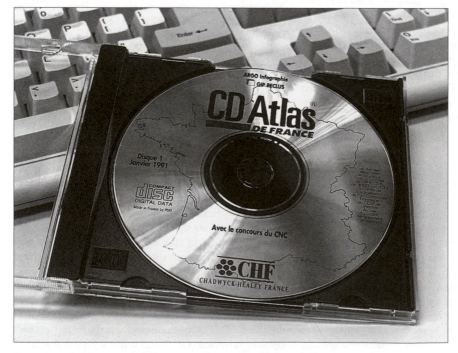

Figure 9.1 CD-ROM Atlas de France (GIP-RECLUS, 1991)

In *analytical electronic atlases* the full potential of the electronic environment is used. Apart from the map object query function described above for the NAIS, data sets can be combined, so that the atlas user is no longer restricted by the themes selected by the cartographer for the atlas. Computations can be effectuated on areas, on themes, or on themes within specifically defined boundaries, and much of the GIS functionality would be available here. Still, the major emphasis will be on assessing the spatial information and on the visualization of the result. If another definition of electronic atlases is presented here, it is to this third, analytical type that it applies mainly (Van Elzakker, 1993): 'An electronic atlas is a computerized GIS, related to a certain area or theme in connection with a given purpose, with an additional narrative faculty in which maps play a dominant role.'

Though the access to electronic atlases is of course limited to PC locations, they provide a number of advantages over paper atlases worth noting:

- Customized maps can be produced on them. The InfoNation atlas (Electromap, USA) is a good example: in a special legend window all object categories one wants to include in the map can be clicked on, and an inset detail map will show immediately the effect of this decision. Names can be added at will for each object category (Figure 9.2).
- Geometrical information can be provided immediately: distances between points, lengths of routes, areas of regions, however defined, are provided upon request, as are the geographical coordinates of the cursor position.
- Whereas traditional atlases isolate (they only show a particular area, at a particular scale, pertaining to a specific date or period in time), electronic

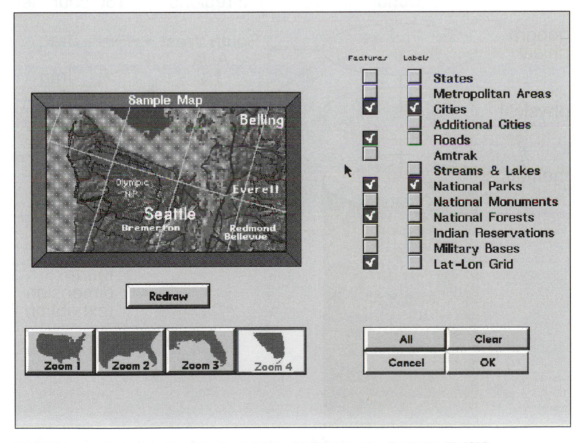

Figure 9.2 Legend category selection window from InfoNation 2.0 (Electromap Inc., Fayetteville, AR, 1993).

atlases have the power to shed these restrictions, and to move beyond map frames by panning, zooming and presenting animations that show developments over time.

- Highlighting or clicking a name in the register or index will immediately provide a map showing the named object on the largest scale available. On the other hand, clicking on an object on the map will provide its name: Bartholomew's Times World Map and Database is a prime example of this functionality. So there need not be a clutter of names on the map.

- In more recent versions of electronic atlases (like Software Toolwork's World Atlas) the link with multimedia has been forged. Icons will show the objects of which photographs or sound tracks have been stored as well; photographs are now already being replaced by videoclips.

- The new French national atlas allows one to aggregate thematic data to larger regions, and this is an important option for map use. New patterns are found as the scale of the enumeration area changes. Soon it will be possible to effectuate this with a slide bar.

Two aspects of electronic atlases maintain the same overall importance as in traditional atlases: one is access and navigation; the other is the ability to compare.

An electronic atlas (and indeed all atlases) is only useful if its users have a clear idea of its overall possibilities and structure, of the way to access the information they want and of the way to get back to the starting point. In order to realize this, they must have 'maps' of these electronic atlases at their disposal (Figure 9.3), and the atlas should have a function showing where its users are on this map.

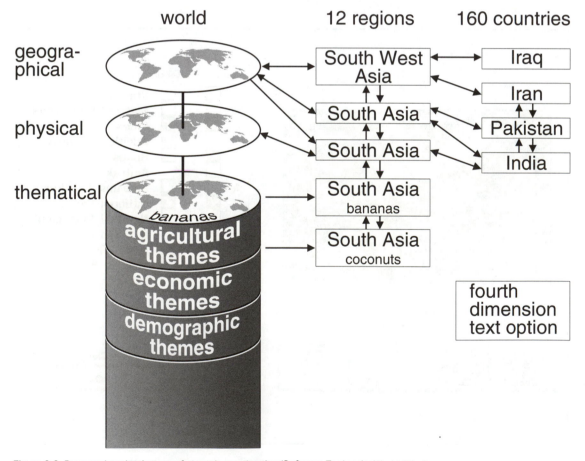

Figure 9.3 Proposed navigation map for an electronic atlas (Software Toolworks World Atlas)

The ability to compare maps is one of the essential characteristics of atlases: by processing the spatio-temporal data, atlas editors see to it that individual maps can be fruitfully compared to the other maps contained. These comparisons can be of a thematical/topical nature (e.g. comparing illiteracy and average income for the countries of a continent), of a geographical nature (e.g. comparing settlement patterns in the United States with those in China) or of a temporal nature (comparing land use in Israel between 1910 and 1970) (see also Section 4.3). To be fruitful the documents compared must be on the same scale, with similarly detailed base maps available, generalized in a similar manner. The settlement densities of areas should reflect, if possible, population densities. For relevant comparisons all maps at a specific scale must have been drawn by applying the same generalization rules. It is only relevant to compare maps when the representation on these maps has indeed been standardized. Apart from generalization, this standardization covers the use of symbols, representational values and scale series.

If accessibility and comparability have been accounted for, other aspects can be taken into account, and amongst these the specific atlas objective comes foremost, as expressed by the existence of different atlas types. In the past, a number of differ-

ent types of traditional paper atlases have developed. One might discern between reference atlases, school atlases, topographic atlases, topical atlases (which only represent one particular theme for many areas, such as the *FAO World Atlas of Agriculture* or the *Atlas of War and Peace*) and national atlases. This last category can be defined as: atlases that contain a comprehensive combination of high-resolution geographical data sets that each completely cover the same country. Now, in the electronic environment, a similar differentiation of atlas types is taking place. National atlases were the first to be developed into national atlas information systems, and reference atlases followed. It will be only a matter of time before other electronic atlas types emerge, such as earth sciences atlas information systems, physical planning atlas information systems, socio-economic atlas information systems or historical atlas information systems.

As a matter of fact, there already exists a good example of the latter, i.e. Millennium (Figure 9.4). This is an electronic atlas showing the changes in the political history of Europe since AD1000. The spatial and temporal (with a monthly increment) data on every boundary change have been stored in the file and each of them can be visualized. The program can be put in a motion-mode, which will result

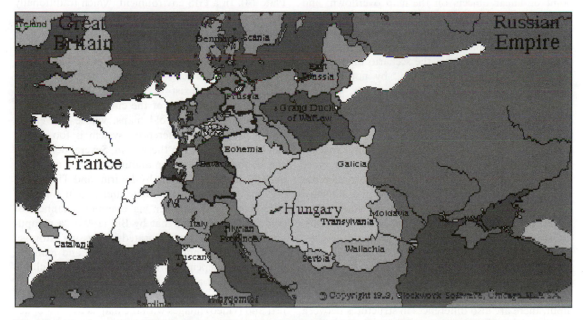

Figure 9.4 Napoleon's incursion into Russia, derived from Millenium 1.21 Clockwork Software Chicago (1993)

in an animated picture of the historical developments, such as Napoleon's incursions into Russia. Textual descriptions of the events are provided, of course, and it will be only a matter of time before other elements, such as imagery or even excerpts from historical motion pictures, are added as well.

The overall objectives of the various types of atlas information systems can be translated in the relation of each data set contained to the other data sets. These relationships are generally elucidated through the use of metaphors and materialize in the form of specific structures and scenarios. These scenarios form the starting point for the design of graphical user interfaces (GUIs). The objective of an electronic school atlas of Sweden, for example, would be to communicate basic spatial information about the country in such a way that it would keep pupils interested. In order to effectuate that, the scenario could be a simulation of the flight of geese from south to north over the country, thus providing a gradual overview (as the user would see it from above) of the country's geography, with points indicated where adventures can be had (to keep up the interest), but also allowing for free roaming in order to discover patterns on one's own. So the GUI would have to provide functions for taking off and for landing again, and for changing weather conditions. The structure of the data would have to allow for access at all points, for providing information about all point or linear objects on the map overflown, and for links with other media at specified locations.

The various electronic atlas types discerned will each have different scenarios and structures. For the strictly reference type, the scenario might be to simulate a complete impartiality, by not pre-programming any Eurocentric viewpoints, or, in order to boost the user's interest, it might be directed at comparisons between different regions, and to quantify these comparisons. Socio-economic atlas information systems should have an in-built capacity, for instance, to numerically compare data for other regions to that of the user's home region (or reference region), so that the user can find out in a standard way whether people in other areas are worse off or not. For physical planning atlas information systems, the scenarios should simulate the planning process, so that the user gets the feeling of participation (as provided in the SimCity computer game), instead of just being a bystander.

As the scenarios would allow for more or less user input, there are also differences in structures between these various types of atlas information systems.

National atlas information systems would focus on one country, and only occasionally compare the situation in this one country to the wider world. The opposite is valid for reference atlases which would strive to contain a specific level of detailed topographic/chorographic information for the whole area covered.

Apart from the commercially or otherwise produced ready-made atlases, authoring systems or atlas shells have now been made available (Smith and Parker, 1995) that can be used as containers to put atlas data in, and then provide all the functionality to access the information (at least in the way interactive atlases work). In this way, electronic atlases can be produced with very low costs and with a substantial gain in time.

Electronic atlases are now increasingly linked with multimedia systems, through hypertext structures (see Section 9.3), so there is an effortless hopping between maps (both static and animated), graphs, text and illustrations (both stills and animations).

9.3 Maps and multimedia systems

Multimedia allows for interactive integration of sound, animations, text and (video) images (see Plate 14). In a GIS environment, which traditionally can work only with coordinates, pixels, their attributes and spatial relations, this new technology offers a link, often via the map, to all kinds of other information of a geographic nature. These could be text documents describing a parcel, photographs of objects that do exist in the GIS database, or a video of the landscape of the current study area. It could also mean that old maps, whose conversion into a properly georeferenced system is too costly, can be incorporated in the system. Several definitions can be found to describe multimedia. The most frequently cited are the one by Laurini and Thompson (1992): 'a variety of analogue and digital forms of data that come together via common channels of communication', and that by Bill (1994): 'a computer-based system for integrated processing, storage, presentation, communication, creation and manipulation of independent information from multiple time-dependent and time-independent media.'

The objective of combining sound, animations, text and (video) images with the map is to get a better understanding of the mapped phenomena as a whole.

Figure 9.5 Multimedia components and the map

Figure 9.5 schematically demonstrates possible relations between multimedia components and the map. It is also possible to use the map as a kind of index to phenomena or objects that are represented by one of the multimedia components. Since multimedia equipment can produce music of CD quality one can easily imagine what a multimedia composers' map of Europe would look like. Pointing to Beethoven's place of birth would activate his ninth symphony, show his picture and music sheets as well as a video of a landscape that visualizes the atmosphere of the music. This level of integration is almost realized in some of today's interactive multimedia encyclopedias (see, for instance, Compton's Interactive Encyclopedia 1995). But what is the status of Multimedia within GIS and cartography?

Today's GIS packages have only limited capacity to handle (except for a bleep to signal an error), video and animations. Text can be generated from the database and often scanned images of photographs of geographic objects, of paper maps and of text documents can be displayed. Currently the GIS packages are getting better integrated in the desktop environment, as explained in the previous chapter. The user is guided by a generic graphical user interface that allows the display of maps, and provides access to the data behind these maps. Pointing at an object on the map would immediately highlight the corresponding record in the database or diagram. In the same way they would open up toward multimedia and the general desktop environment. The map allows for direct links with spreadsheets, and video and animation. If we want to incorporate this new technology into our spatial data handling systems, questions have to be answered. Amongst these are how to structure multimedia information (in order to allow for sound navigation strategies), how multimedia can be used to enhance spatial analysis, what the role of the map will be, and what interface design will involve.

The following sections present the relation between the map and the individual multimedia components (sound, text, (video) images and animation) in relation to visual exploration, analysis and presentation.

9.3.1 Sound

Maps supported with sound to present spatial information are often less interactive than those used to analyse or explore. Examples are maps as indexes to sound libraries. In some electronic atlases pointing at a country on a world map would initiate the national anthem of the particular country being played, or some general phrases in the country's language being spoken (Mindscape's World Atlas 5.0, 1994). In this category one can also find the application of sound as background music to enhance a mapped phenomena, such as industry, infrastructure or history. Experiments with maps in relation to sound are known on topics such as noise nuisance and map accuracy (Fisher, 1994; Krygier, 1994). In both cases the volume of noise is controlled by pointing at a location on the map. Moving the pointer to a less accurately mapped region would increase the noise level. Both are examples related to analysis. The same approach could be used to explore a country's language. Moving the mouse to a particular region would start a short sentence in the region's dialect. Krygier (1994) has experimented with sound as an additional variable to graphical variables like colour and size. In virtual reality environments three-dimensional sound is used for orientation purposes.

9.3.2   Text

GIS is probably the best representation of the link between a map and text (the GIS database). Imagine a map showing a country's population density, in which all provinces are coloured according to one of the four different classes discerned. To amplify the presentation the user can point to a province, which will result in the display of its name and its actual individual population density value. Electronic atlases often have all kinds of encylopedic information linked to the map as a whole or to individual map elements. It is possible to analyse or explore this information. Country statistics can be compared. Clicking Lisbon on the map of Portugal would reveal a list with the most important tourist sights, or even activate other multimedia components such as starting a video tour of the city.

9.3.3   (Video) images

Maps are models of reality. Linking video or photographs to the map will offer the user a different view of reality. Topographic maps present the landscape. Next to this map a non-interpreted satellite image or aerial photograph can help users in their understanding of the landscape. While analysing a geological map it can be enhanced by showing landscape views (video or photographs) of characteristic spots in the area. A real estate agent could use the map as an index to explore all the properties he or she has for sale. Pointing at a specific house would show a photograph of the house and the construction drawings, and a video would start showing the house's interior.

9.3.4   Animations

Maps often represent complex processes. Animations can be very expressive in explaining these processes (see also Section 9.4). To present, for instance, the structure of a city they can be used to show subsequent map layers which explain the logic of this structure (first relief, followed by hydrography, infrastructure, and land use, etc.). Animations are also an excellent means to introduce spatial data's temporal component: the evolution of a river delta, the history of the Netherlands coastline or the weather conditions of last week.

From a technical point of view, there are almost no barriers left. The user is confronted with a screen with multiple windows displaying text, maps, even video images supported by sound. The important issue is to keep in control over all the options, and to allow the users to manage all the information that will reach them. It is most important, therefore, to state the objectives or the purpose of use (exploration, analysis or presentation) in advance, so that there will be a yardstick to measure the result by.

9.3.5   Hypermaps

In the examples in the previous section the map was used to link and order the individual multimedia components. For users of spatial data a map is also a natural access medium or interface to the data. To help solve the structuring and interface problems discussed above, this map approach should be elaborated upon. This introduces the hypermap concept. From a visualization perspective a hypermap can play a key role in structuring the individual multimedia components in respect to each other and to the map, and will allow the user to navigate the date. It can be described as georeferenced multimedia. This concept, introduced by Laurini (Laurini and Thompson, 1992) is based on the hypertext and hyperdocument principles. Hypertext can be described as a set of nodes (abstractions of text or graphics) which are connected by links that allow the user a non-sequential tour around the data. Apple's HyperCard made this principle widely available. Hyperdocuments introduce multimedia components into the hypercard concept. Hypermaps introduce spatial referencing to all components in the system and allow for a

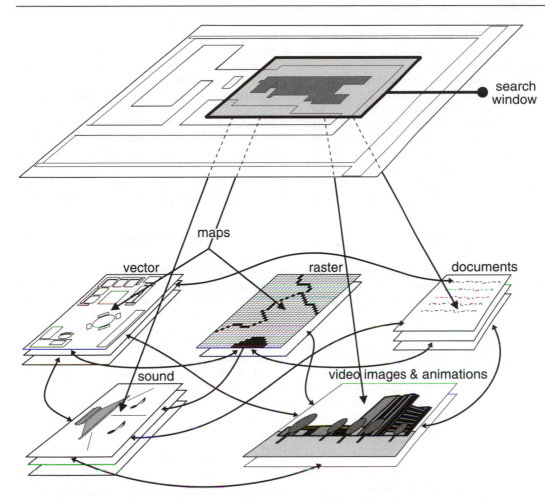

Figure 9.6 Hypermap principles: finding multimedia documents and their links based on a spatial search

spatial and thematic navigation around the data. All possible links are predetermined, but individual users do not necessarily follow the same paths. An example illustrates its use. Identifying a university complex on the map (Figure 9.6) could give access to more detailed maps and plans of the area's sewerage system, individual buildings, their photographs or a video of the daily life in the building. It would also be possible to get an overview of the nature of the lectures given at that location. For example, for the foreign language department, the notes for a particular course could be read, or heard. Pointing at the lecture room where the course is given would return the user to the map. Spatial and thematic links would be provided all over, with the map as a starting point.

9.4   Animated maps

The need in the GIS environment to deal with processes as a whole, and no longer with single time-slices also influences the visualization aspects. In order to visualize models or plan operations, animated maps are becoming more efficient than static paper maps. On-screen maps offer opportunities to work with moving and blinking symbols, and are most suitable for animations. Next to some animation basics this section will address two topics relevant to animations. The first is their application. When does it make sense to use an animation to visualize spatial data? Should it only be used when temporal change is at hand or can it be used to clarify non-temporal spatial relations as well? The second issue relates to user interaction and visualization circumstances. The user should be able to interact with the animation, otherwise they do not add much to a series of static maps of the same topic. The functionality of the interface will depend on the type of use, such as exploration or presentation. Animation is the process of design and production of images that suggest movement. It is a strong method of visual communication, especially because it can deal with real data, as well as abstract and conceptual data.

Computer animations can be generated according to three methods:

- *Frame by frame*. This approach is the most elementary animation method. Each single image (or frame) is created separately, and finally all frames are combined into an animation. To suggest continuous movement one needs between 24 and 30 frames per second.
- *Key frame*. Here only the most characteristic images are created (the key frames), and a computer program will interpolate those frames in between two key frames.
- *Algorithmic animation*. This is currently the most powerful animation technique. A computer program defines what will happen during the animation. The creator will only define objects, changes and the moments changes take place.

Animations can be displayed off-line or on-line. In the off-line situation one applies analogue video techniques to record the individual frames. On-line animation can be done as real-time animation, e.g. all calculations to create the individual frames are done directly from the database and displayed simultaneously. This is the most interesting approach for a GIS environment. Animations in electronic atlases are often of a 'real-time later' animation type (i.e. their computation has been effectuated in advance): all individual frames are stored in a file, from which the animation is played.

Cartographers have paid attention to animation since the 1960s. However, early work only allowed for the non-digital cartoon approach, and experiments were made with either film or television. During the 1980s technological developments gave rise to a second phase of cartographic animation, with the first computer-produced imagery. Currently a third wave of cartographic animation is going on, created and enabled by GIS technology. A historic overview is given by Campbell and Egbert

(1990). The method currently most often applied in cartography is the frame-by-frame approach.

As mentioned before, animations can be very useful in clarifying trends and processes, as well as in explaining or providing insight into spatial relations. It is possible to categorize cartographic animations in temporal and non-temporal animations. To understand this classification the notion 'display time' has to be introduced. Display time can be described as the moment the viewer of an animation actually sees the images. A direct relation does exist between the individual frames and the display time (Figure 9.7).

9.4.1   Temporal animations

When dealing with a temporal animation, a direct relation also exists between display time and world time. World time is the time-scale of reality, i.e. the moment an event takes place in the real world. Examples of these animations are changes in the Netherlands coastline from Roman times until today, boundary changes in Africa since the Second World War, or the changes in yesterday's weather. Time units can be seconds, years or millennia. The GIS environment also distinguishes another type of time, database time, i.e. the moment a real world event is registered in the database. These three different types of time were already recognized, although not explicitly, by the cartographer working on topographic maps. Topographic map updates would be a good example, as here a difference of several years would exist between world time, database time and display time (respectively the moment a new road is built, the areal photograph taken, and the final map is printed). Temporal animations show change, changes in the locational or attribute components of spatial data, as shown in Figure 9.8(a). It is important that the user can influence the flow of the animation. He or she should be able to play with the time line: forward, backwards, slow, fast, pause.

9.4.2   Non-temporal animations

Display time in non-temporal animations is not directly linked with world time. The dynamics of the map are used to show spatial relations or to clarify geometrical or attribute characteristics of spatial phenomena. Here interaction is necessary as well, if

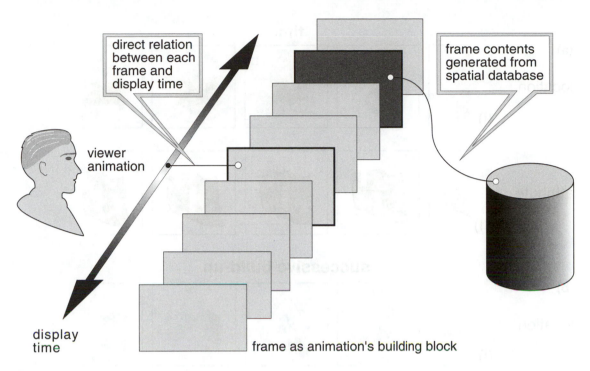

direct relation
between each
frame and
display time

frame contents
generated from
spatial database

viewer
animation

display
time

frame as animation's building block

Figure 9.7 Animation frames and display time

only to allow the user to answer the question 'How was it . . . ?' Non-temporal animations can be split into those displaying a successive buildup of phenomena, and those showing changing representations of the same phenomena (Figures 9.8b and c).

Examples of animations with successive buildup include the following:

- Understanding a three-dimensional landscape. For instance, first only the terrain is displayed, followed later by the addition of other themes such as roads, land use and hydrography (location).
- In thematic mapping alternating classes are highlighted to show, for instance, the distribution of low and high values (attributes).

Animations with changing representations (different data or graphic manipulations) which is in fact the same as the toggling mechanism described for electronic atlases (Sections 9.2 and 10.2), include the following:

- a display of choropleths with different classification methods used (attribute);
- displaying a particular data set by changing the cartographic method of representation, for

instance by showing the same data subsequently in a dot map, a choropleth, a stepped statistical surface and an isoline map (location/attribute);
- maps with blinking symbols to attract attention to a certain location on the map (attributes);
- a simulated flight through the landscape, as a result of continuous changes in the viewpoint of the user (location);
- the effects of panning and zooming in or out in animation (location and attribute).

9.4.3 Dynamic variables

A question cartographers have to deal with is how one can design an animation to make sure the viewer indeed understands the trend or phenomena. The traditional graphic variables, as explained in Chapter 6, are used to represent the spatial data in each individual frame. Bertin, the first to write on graphical variables, had a negative approach to dynamic maps. He stated (Bertin, 1983), 'movement only introduces one additional variable, it will be dominant, it will distract all attention from the other (graphical) variables'. Recent research, how-

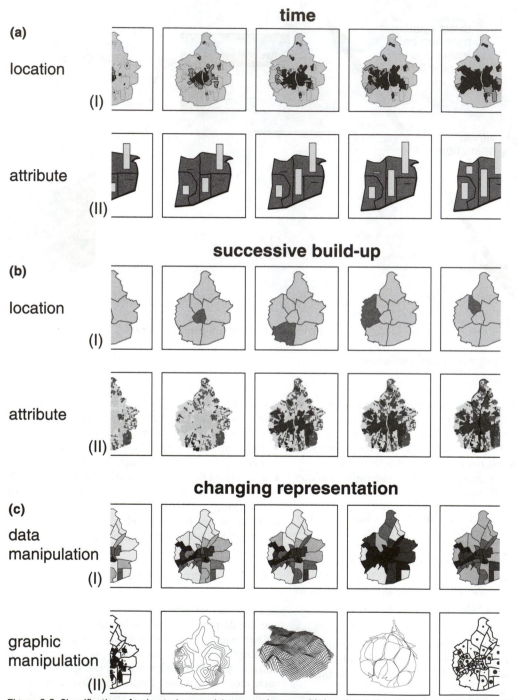

Figure 9.8 Classification of animated maps: (a) temporal maps with locational (I) and attribute (II) change; (b) successive buildup according to location (I) or attitude (II); (c) changing representations because of data manipulation (I) or graphic change (II)

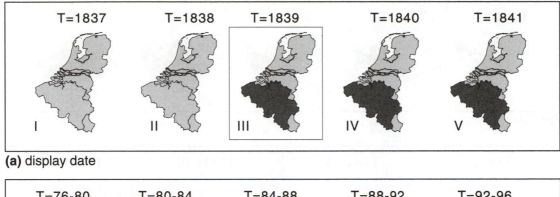

(a) display date

(b) duration

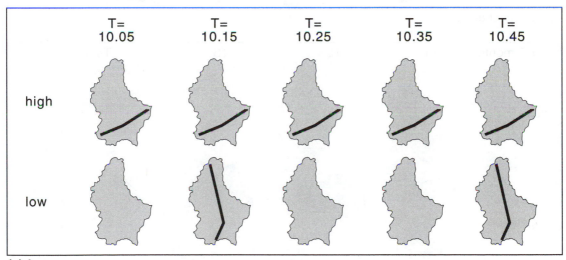

(c) frequency

Figure 9.9 Dynamic variables: (a) display time; (b) duration; (c) frequency; (after Kraak and MacEachren, 1994)

ever, has not sustained this statement (Ormeling *et al.*, 1996). Here we should remember that technological opportunities offered at the end of the 1960s were limited compared with those of today. DiBiase *et al.* (1992) found that movement would reinforce the traditional graphical variables.

In this framework DiBiase introduced three so-called dynamic variables: duration, order and rate of change. In 1994, MacEachren added frequency, display time and synchronization to this list. The characteristics of the dynamic variables are given below and illustrated in Figure 9.8 with examples

(d) order

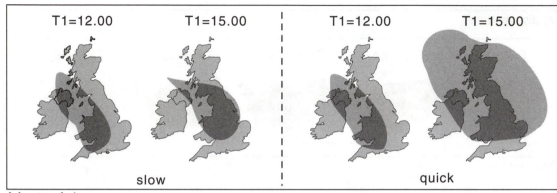

(e) rate of change

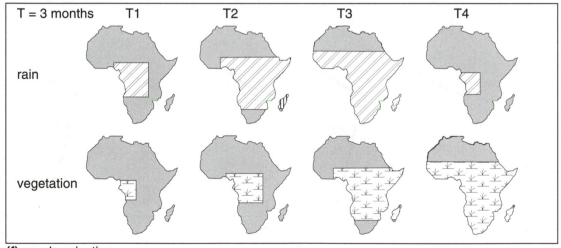

(f) synchronization

Figure 9.9 Dynamic variables: (d) order; (e) rate of change; (f) synchronization (after Kraak and MacEachren, 1994)

for temporal and non-temporal animations (Kraak and MacEachren, 1994).

- *Display time.* This is the time at which some display change is initiated. In Figure 9.9(a)I this is the moment a symbol appears in the Benelux map. The

display date can be linked directly to the chronological date to define a temporal location. In Figure 9.9(a)II this is illustrated by the moment Belgium changes colour when it becomes independent.

- *Duration*. The length of time nothing changes in the display. In Figure 9.9(b)I the United States is highlighted for several seconds. The temporal example in Figure 9.9(b) shows how long a political party has been in office in the United States. A direct link between each frame and world time exists.
- *Frequency*. Frequency is linked with duration. Either can be defined in terms of the other. It is worth treating it as a separate dynamic variable because humans react to frequency as if it were an independent variable. In Figure 9.9(c), frequency is linked to temporal data, the train timetable in Luxemburg.
- *Order*. This refers to the sequence of frames or scenes. Time is inherently ordered. Figure 9.9(d) shows how a phenomenon spread over the United States over time. Non-temporal order is shown by highlighting the different regions in the United States one after the other.
- *Rate of change*. The difference in magnitude of change per unit time for each of a sequence of frames or scenes. Change can be in location or in attributes at locations. On a dynamic map, with time controlled via a fixed scene length, both location and attributes can change, and they can change at different rates. The rate of change can be constant or variable. It will, of course, be zero if there is no change. If neither the magnitude nor the duration of change is zero, either or both can be controlled to produce an increasing, constant or decreasing rate of change. Figure 9.9(e)I shows slow changes and Figure 9.9(e)II shows quick changes in the cloud cover over Britain.
- *Synchronization*. Synchronization (phase correspondence) refers to the temporal correspondence of two or more time series. Only temporal examples exist (Figure 9.9f). It can apply to matching chronological data of two or more data sets precisely. If natural patterns are out of phase (such as rainfall and vegetation growth in Africa), adjusting their synchronization at the display stage may uncover causal links that would not be apparent if chronological dates are closely adhered to.

These dynamic variables can be seen as additional tools to design an animation. They can control all visual manipulations. Animations have a narrative character. They tell a story. The variables' duration and order have an especially strong impact on this story. The order in which the images are presented, and the duration of each of the images will determine

the message conveyed. It is also possible to use the dynamic variables directly to represent spatial data. An example of this use is the use of the variable duration, to show uncertainty. Those parts of the map with high certainty are stable and those with a low certainty appear chaotic by blinking.

It has been mentioned that user interaction with the animation is essential. A question to be asked is how much interaction a user should have. As will be explained in more detail in the next section, each view environment requires its own interface. It will be clear that a viewer in an exploratory environment does not want to be limited by interface tools, especially because in this situation it is very likely that he/she is the author of the animation as well. In a presentation environment the interface tools can be limited. The order of the frames has been set by the author, and facilities to go forwards, backwards, at different speeds, to jump from one scene to another, and to stop or pause, will suffice.

9.5 Scientific visualization and exploratory data analysis

Before the GIS era, paper maps and statistics were probably the most prominent tools for researchers to study their spatial data. Wood (1994) provides a well-referenced overview of the important role paper maps played in spatial studies. His work, which places the map in the centre of a wider context, confirms this. To work with those paper maps, analytical and map use techniques were developed, which can still be found behind many GIS packages' commands. Today the same researchers have available large powerful sets of computerized tools like spreadsheets and databases, as well as graphic tools, to support their investigations. Comparing the on-screen approach with the traditional approach not only reveals a difference in processing effort and time (computers do in seconds what took weeks using manual methods), but now the user can interact with the map and the data behind it. This puts the map in a different perspective, and more than ever before at the centre of spatial studies. This development has led to the introduction of the term cartographic visualization.

The term visualization should be explained in more detail. Some would wonder why, since cartography is equal to 'making visible', visualization is differentiated from cartography (see also Section 3.1). Is this not what cartographers have been doing for centuries? However, since a publication by McCormick *et al.* (1987). the term scientific visualization has been used in a wider context. On-screen graphics were seen to be of major importance to stimulate and enhance insight in data behind the graphics. In a research context the term scientific visualization was defined as 'the use of sophisticated computer technology to create visual displays, the goal of which is to facilitate thinking and problem solving'. Interaction with the data plays a key role in scientific visualization. In 1994, two books, *Visualization in Geographical Information Systems* (Hearnshaw and Unwin, 1994) and *Visualization in Modern Cartography* (MacEachren and Taylor, 1994) addressed the relations between the fields of cartography and GIS on one hand, and scientific visualization on the other, in more depth. In the international cartographic community it has been accepted that visualization is not just equal to cartography. It is seen as an independent development that will have major influence on cartography (Taylor, 1994). In his view the basic aspects (Figure 9.10), cognition (analysis and application), communication (new display techniques) and formalism (new computer technologies) are linked by interactive visualization.

Scientific visualization is seen as a three-stage process including filtering and enhancement of the data, mapping the data (meant in the mathematical sense), and rendering the image. This visualization process is schematically shown in Figure 9.11, where it is placed next to the cartographic process. The first phase shows that the model resulting from the data filtering and enhancement steps can be compared with the Digital Landscape Model. In a cartographic context the visualization techniques will result in a Digital Cartographic Model. The rendering process will result in a (map) image. As well as these similarities there are also some clear differences. In the scientific visualization process, as in cognitive cartography, emphasis is on exploration of the data to gain understanding and insight, and pay less attention to communicative aspects of the image. Aspects which worry cartographers, such as the nature and origin of the data (accuracy, currency and completeness), get less attention (Figure 1.15).

However, it has been recognized that data exploration techniques are gaining importance, especially as GISs move from a data-poor to a data-rich environment. This was explained in Chapter 1, where the terms visual communication and visual thinking

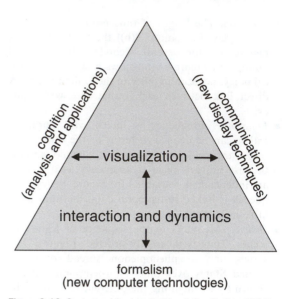

Figure 9.10 Cartographic visualization (after Taylor, 1994)

were introduced (see again Figures 1.1 and 1.2). It was developments in scientific visualization that inspired cartographers to introduce this terminology in cartography. Methods such as the cartographic grammar (Chapter 6) that exist for communication cartography are not yet available for analytical cartography. In the literature several experiments with cartographic exploratory data analysis environments have been described. MacDougall (1992) has developed a prototype of such a system, Polygon Explorer (Figure 9.12a). It shows a map of Massachusetts in relation with some diagrams after specific queries have been executed. Another example is the geographic brushing technique introduced by Monmonier (1992). As Figure 9.12(b) shows, links exist between a map and several scatterplots. Selection of a set of values in one scatterplot will highlight the corresponding areas in the map and the points in the other scatterplots. Conversely one can also select areas in the map and see their corresponding points highlighted in the scatterplots. DiBiase *et al.* (1992) give an overview of other approaches as well.

Further reading

Bertin, J. (1983) *Semiology of graphics*. Madison, WI: Wisconsin University Press.

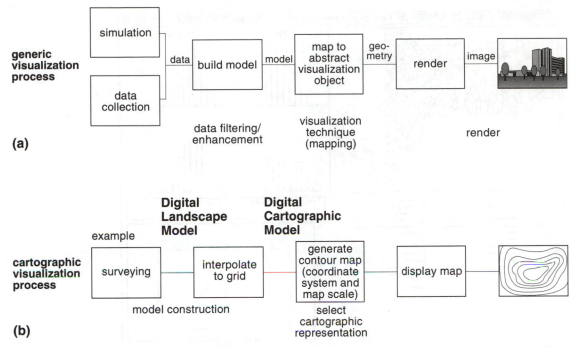

(a)

(b)

Figure 9.11 The visualization pipeline: (a) generic visualization process; (b) cartographic visualization process (adapted from Brodlie, 1994)

Buttenfield, B.P. and R.B. McMaster (eds) (1991) *Map generalization: making decisions for knowledge representation.* London: Longman.

Campbell, C.S. and S.L. Egbert (1990) Animated cartography/Thirty years of scratching the surface. *Cartographica,* **27**(2), 24–26.

Hearnshaw, H.M. and D.J. Unwin (1994) *Visualization in geographical information systems.* London: Wiley.

MacEachren, A.M. and D.R.F. Taylor (1994) *Visualization in modern cartography.* Modern Cartography series. Oxford: Pergamon.

McCormick, B., T. DeFanti and M.D. Brown (1987) Visualization in scientific computing. *Computer Graphics,* **21**(6).

Products mentioned

CD-ROM Atlas de France. 1991 GIP/RECLUS Montpelier.

*Compton's Interactive Encylopedia for Windows. 1995. Compton's New Media Inc.

InfoNation. 1991 Electromap, Fayetteville.

Millennium 1.21. 1993 Clockwork Software. Chicago.

Times World Map and Database. 1994. London: Times Books.

SimCity. Maxis 1992.

World Atlas. 1994 Mindscape's

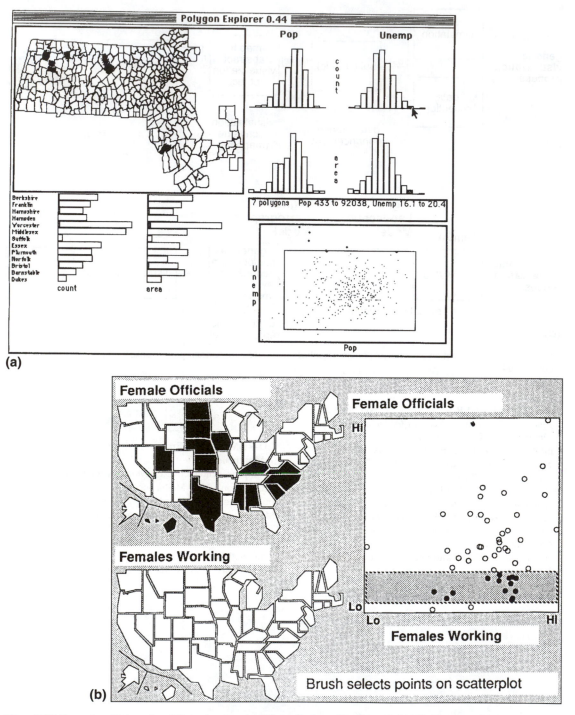

Figure 9.12 Examples of interactive data exploration: (a) Polygon Explorer (MacDougall, 1992); (b) geographic brushing (from Monmonier, 1992)

CHAPTER 10

Cartography at work: maps as decision tools

10.1 Again: why maps?

In the past nine chapters examples have been provided that illustrate what maps can do in analysing and communicating spatial information. By providing them, readers will have become familiar with many aspects of Britain's Lake District or of Maastricht Municipality in the Netherlands. Without these maps it would have been difficult to decide on answers to the questions or problems stated, it would have been difficult to decide upon a course of research, or to understand the spatial impact of environmental factors. Maps help one in deciding what to analyse, and later on they support one in formulating decisions on issues with a spatial impact and in communicating these decisions.

10.1.1 Explaining patterns

By juxtaposing the map of a specific pattern with other maps of the same area, with different map themes, correlations might be found, and these might help us to find causal relationships. The relief map of the Kilimanjaro region in Figure 5.26 has been taken from the Digital Chart of the World. When this area is printed out with its highway infrastructure and settlements (Figure 10.1a), a discrepancy between the southeastern and northwestern part of this mountain area emerges. This discrepancy can be explained by showing the extent of agricultural land, as, indeed, it is only on the southeastern side of the mountain that an extensive area is used as arable land (Figure 10.1b). The extent of the arable land can be explained, theoretically, by local soil patterns, slopes, heights and precipitation. As the first three factors are similar all around the mountain, it must be due to the influence of differences in precipitation and indeed, when one compares these two maps with a precipitation map (Figure 10.1c), a strong correlation between the infrastructure and agricultural patterns and the rainfall pattern emerges.

10.1.2 Comparison and analysis

In similar ways, other patterns presented before might be explained: the differences in traffic accidents as shown in Figure 4.6 between the municipalities of Middelstum and Hoogezand-Sappemeer can be explained by juxtaposing them with number of vehicles maps, population distribution maps (Figure 10.2b), road maps (Figure 10.2a), length of road maps or vehicle per kilometre of roads per municipality maps of the same area: if all other conditions are similar, municipalities with more inhabitants would be likely to have more traffic accidents; similarly, municipalities with motorway entries and exits and a larger overall road length, more cars or more cars per standard length of road would be liable to have more accidents. It would depend to some extent on the local road safety conditions (as the Netherlands is well known for its multitudes of undisciplined cyclists, separate bicycle paths would make a lot of difference), differences in alcohol consumption and general attitudes.

Differences in population numbers alone can never explain differences in accident numbers completely: Figure 4.17 shows that with only about 73 km^2, Barrow district has a population density of nearly 1000 persons per km^2, and is therefore pretty much urbanized. This would bring a higher chance of public transport, shorter distances to shops and malls,

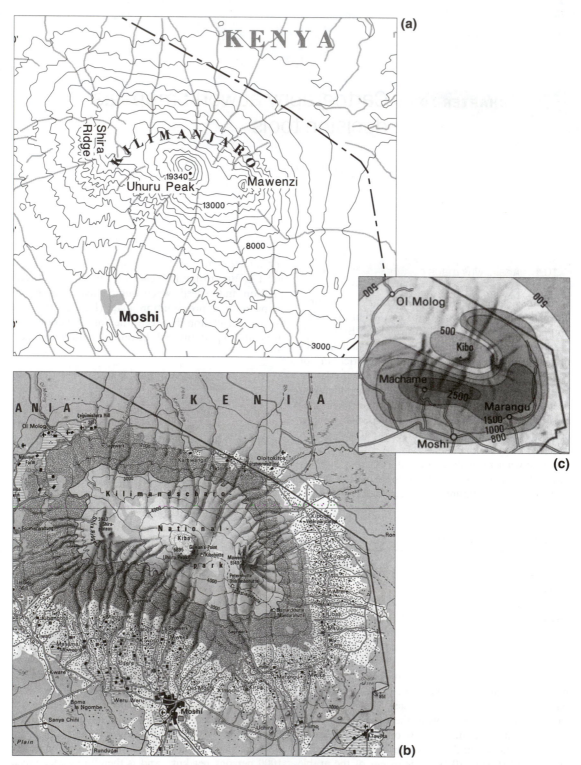

Figure 10.1 Understanding the Kilimanjaro region: (a) general topography concentrated on terrain shapes (derived from Digital Chart of the World); (b) land use map; and (c) precipitation map (both derived from *Alexander WeltAtlas*, 1989). The lightest area in which most settlements are located is the arable land area

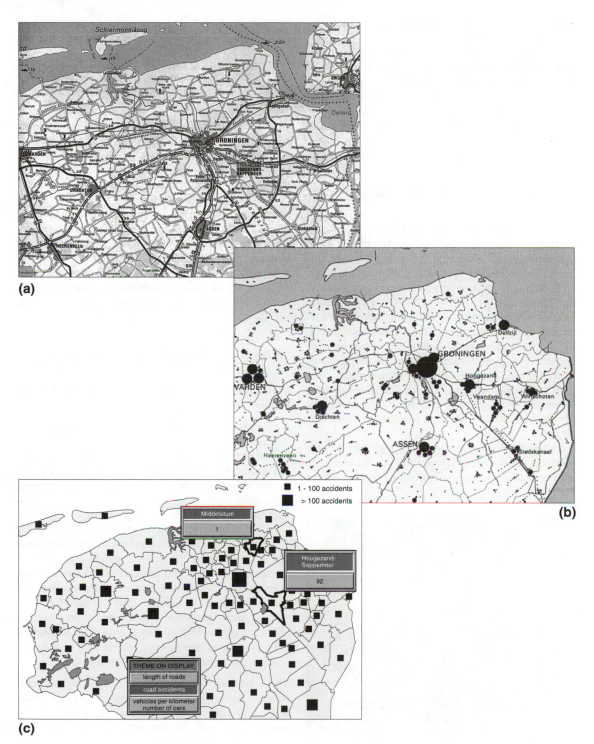

Figure 10.2 Northern Netherlands: (a) a road map of the area (derived from ANWB-road map 1993), (b) population distribution (derived from *National Atlas of the Netherlands*, 2nd edition, volume I; (c) number of road accidents (see Figure 4.6)

and therefore a lower percentage of car owners than in other districts of Cumbria.

As soon as such hypotheses have been formulated, proof must be found, and the juxtaposition of maps is a good way of ascertaining the factors that will probably explain the differences, and can be tested for their relevance for the issue statistically. Of course there needs to be the facility to juxtapose the maps, and this calls for a facility to split the monitor screen, and show at least two maps simultaneously.

10.1.3 Analysis and decision making

Figure 10.3 shows the centre of Maastricht, with an overprint of newspaper distribution areas (newspapers are delivered to subscribers in the Netherlands) and of the number of subscribers per area. When these numbers have been compared to the actual numbers of households per distribution area, it will be possible to assess the degree of penetration of the newspaper. Those areas with a relatively low percentage of subscribers could then be targeted by advertisements or special campaigns or offers for trial subscriptions at bargain prices – but such campaigns will only be sensible in those areas where the average socio-economic level would be compatible with the message brought by this newspaper.

The next step would then be to highlight the distribution areas for which positive action has been decided, where bargain subscriptions will be offered

to the non-subscribers, and to circulate these maps to those working in the newspaper's marketing department.

10.1.4 Maps as interfaces with databases

In all these examples and also in those presented in other chapters, maps have also been used as graphical user interfaces to the databanks in which the spatial information was stored. Just by clicking the municipality on the road accident map of the Netherlands, the exact number of victims can be queried (it is never the function of the map itself to provide these exact numbers, because this would distract from the overall patterns and trends it is communicating; for the actual numbers it refers to the database). Other data sets such as length of roads or numbers of cars, or combined parameters such as number of cars per road unit can be accessed just as easily.

10.1.5 Conditions for proper use of the maps

When one wants to profit from this power of maps, certain conditions have to be met:

1 One should be familiar with suitable map-use strategies.
2 One should have access to the relevant data sets.
3 Preferably, meta-information on data quality should be available to assist in the decision-making process.
4 It should be possible to integrate the various data sets, if necessary by modelling them (for instance, generalization); this integration should have been made possible through standardization of exchange formats and geometrical frames of reference. Both the relevant data set (2) and the relevant meta-information (3) can only be found if they have been properly documented. Once again, Figure 10.4 (which is also Figure 1.14) shows the various issues one has to contend with in order to allow for a sensible bout of map use.

These map-use strategies entail identifying the relationship between the mapped objects and their locations and themselves (after having identified the mapped area and the mapped topic, and the way the topic has been encoded in the legend). For this identification a number of map-use components or elements can be discerned: search, locate, identify,

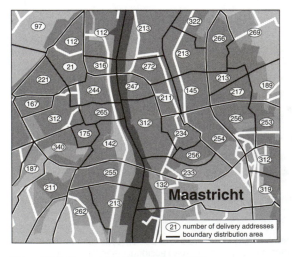

Figure 10.3 Newspaper sales in Maastricht

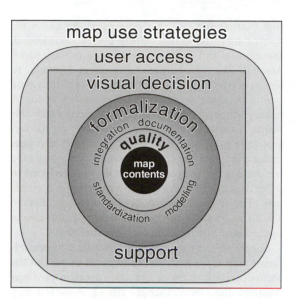

Figure 10.4 Visual decision support for spatio-temporal data handling. Key words in the GIS cartography approach are map use strategies (how people make decisions based on maps), public access (how people work with the information), visual decision support (what the quality of the information is like), formalization (building expert systems). (Based on Kraak *et al.*, 1995)

verify, compare, contrast, delimit, describe, measure (Anson and Ormeling, 1996).

In the case (in Chapter 1) of deciding about the route of the high-speed train in the Netherlands, the map-use task consisted of: *select* maps, *identify* suitable routes, identify restrictions, *select* maps of these restrictions, *identify* and *delineate* restricted areas, *compare* and *add up* or *score* restricted areas, *determine* or *identify* the route with the lowest score, and *decide* on the least noxious route (map tasks indicated in italic).

In the case of the conduit or circuit problem in the utility map example of Chapter 4 (Figure 4.4), the map-use sequence of steps consisted of: *monitor* the map, *detect* the problem, decide on direct action (initiate measures to remove or solve the problem), *determine* consequences (*decide* on the nature of the effects and *delineate* the area being affected), *decide* on the course to mitigate or offset these effects (reroute traffic, for example), *determine* other utilities that might be affected, and *inventorize* equipment in damaged section, etc.

So map tasks can be subdivided into a number of individual actions, during which links or relationships are identified, data are combined, anomalies

are signalled, alternatives are identified, elements are counted or their numbers estimated, sizes are compared, etc. In order to arrive at a specific answer, these individual actions have to be performed in a specific sequence, and the sequence with the smallest number of steps will be the optimal one. It will vary from task to task. A map use task sequence sanctioned by experience is that of navigation: a typical navigation task with a map would consist of: *search* and *locate* one's position on the map, *orient* the map, *search*, *identify* and *locate* one's destination on the map, *determine* options for alternative routes, *select* one of the options, *set* a course, *determine* landmarks by which the course can be identified, *follow* the course on map, *check* landmarks, and *verify* the destination.

10.2   Working with electronic atlases

In an electronic atlas environment these steps in map use are similar, although one is still restricted by the fact that at the moment, two or more maps cannot be visually compared on one monitor screen in most commercially available products. A new element of map use in electronic atlases is the aspect of navigation through the atlas: this gives a new meaning to the '*select* map' task, as the potential map titles will usually be provided in a thematic index. Other new map use tasks are the *clicking* of map objects in order to query them for their attributes and have these presented alphanumerically (such as name of the object, capacity, size, etc.). The *aggregate* option in French electronic atlases refers to the possibility to view the thematic data at other levels of enumeration areas: instead of at the municipality level these data might also be viewed at a district or canton level, a département level or that of an economic region (a grouping of a number of départements). Each new level of presentation would allow one to discover new spatial patterns (see, for example, Figure 3.4, where the high voting percentages for Labour reflect the vicinity of factories or the location of housing blocks developed by denominational building societies at the voting ward level within a city. At a municipality level the high percentage of Labour votes would indicate urban rather than rural environments, while at a provincial level religious denomination patterns would still govern the image.

The same French electronic atlases have the facility to toggle between an absolute and a relative view of the same data set (as in Figure 3.11). With this function it is at least possible to weigh numerically the relative data presented in a choropleth mode and, vice-versa, decide on the relative impact of absolute magnitudes. With an '*add* layer' action, names for example can be added to the map studied, or a more detailed topographic background map. A *highlight* action would lead to the isolation of a specific category or class of objects from its surrounding values or categories, which will be left white or grey. This functionality is contained in Milennium, for example (see Chapter 9). In Figure 10.5 the Hapsburg empire's many possessions in southwestern Germany would pass unnoticed but for this highlighting function, which allows for their presentation in a customized colour, sharply contrasting with a white or grey background.

Other new map use tasks are *panning* and *zooming*, and these refer to changing our window on the world provided by the monitor screen. The fixed, isolating frames of paper maps are exchanged here for flexible boundaries, to be adjusted at will. Though *compare* is a frequently used map use action in a paper map environment, few electronic atlases allow for it yet (in 1995), the electronic national atlas of Sweden being one of the exceptions.

The hypercard-based multimedia environment of current electronic atlases brings with it frequent use of the *hop* command, i.e. the facility to hop to another presentation mode wherever icons or func-tion buttons indicate that possibility. *Browse* is the sort of map use action hardly ever acknowledged, though it is still necessary to provide us with ideas. *Mark* allows one the possibility, as in *highlight*, to simulate the marking with colour pencil/crayon points or areas on the map. The *search* action is usually directed at the electronic index, and this action is matched with the presentation, on the largest scale available, of a map of the object that goes under this name. By the *save*, *download* or *copy* map use actions the images in our files are transferred to other data carriers. *Print* will do the same in a visible way. The *time* and *coordinates* options are used to ask for the (local) time and geographical coordinates of the point at which the cursor is directed.

Toggle will let the map user alternate between different map types such as the physical maps and the socio-economic maps of the same areas in post-war Austrian and German school atlases. *Rank* is a map use action that will list a number of states or regions/cities on the basis of their score for specific topics such as the number of golf courses and/or the length of the yearly period over 18 °C. It can also be used for classifying data into quantiles (e.g. with the top 20% of the observations designated in the legend as 'highest', and the bottom 20% as 'lowest'). *Filter* is the map use action that will set minimum or maximum values to the data presented per enumeration area through proportional circles. Finally, *score* will result in the calculation operations undertaken on the data retained.

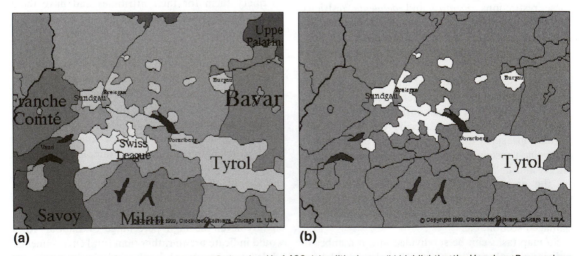

(a)   **(b)**

Figure 10.5 Southern Germany and northern Switzerland in 1400: (a) political map; (b) highlighting the Hapsburg Possessions (derived from Millenium 1.21, Clockwork Software, Chigaco, 1993)

Regarding the map use strategies or map use element sequences, these will typically consist of series of clicks in order for a specific area to be displayed at a given scale, asking for thematic options to be displayed on the monitor, studying them, having them printed and comparing them, in order to assess their similarities and relationships, and taking account of them in providing the answers to the problems stated at the outset.

An approach akin to working with paper atlases would be to browse through electronic atlas maps, marking those maps one would be interested in or think would be helpful in solving a problem, and then asking again for the maps marked out, in a specific sequence. In a multimedia environment this would go together with hopping to tables, imagery, and alphanumeric explanations, whenever deemed necessary or interesting.

10.3 At work with the Digital Chart of the World

GIS data sets are being offered in a big variety today. For small-scale applications the Digital Chart of the World (DCW) at scale 1:1 million is one of the more popular data sources. Many vendors have their own version (i.e. format) of the DCW (such as ESRI, 1993), which was originally published by the Defense Mapping Agency in 1992 (see also Sections 2.4 and 5.6.4). How useful are these data in GIS and cartographic applications? This section will illustrate the potential use of DCW data for three simple spatial analysis operations. If no other data are available, the DCW proves to be a valuable data source. However, some care is needed. The age of the data especially can cause problems. Although the CD-ROMs on which DCW is distributed bear the 1992 imprint, most of the data themselves are older than 1970 (see Figure 2.6). This is especially the case for those areas where the DCW is likely to be the only (digital) source available.

10.3.1 The Netherlands railroads

How suitable is the DCW when one intends to use its data for a network analysis? The railroad layer is one of the many data layers available in DCW. Data for this layer were digitized from existing maps at a scale of 1:1 million. For the Netherlands, ONC E-2, last revised in 1985, was used. The map image contains two line symbols for railroads, one for single and one for double tracks (Figure 10.6a). Those tracks visible on the map have been digitized. Because of the high symbol density on the map, it is sometimes difficult to establish if one is dealing with single or double track. From the map it can also be seen that the tracks stop at the border of builtup areas. To provide a 'connected' network, the gaps in the urban areas have been closed. For several important cities this was done using other data sources (see Amsterdam, Figure 10.6b), and for other cities, such as Utrecht, by adding connectors via a 'spoke of the wheel-like approach'. Figure 10.6(b) shows the varied symbology of the railroad layer, with its 'values' for single, multiple and added connectors. Those familiar with the Netherlands railroads know that most of the country's lines possess double tracks; however, Figure 10.6(b) shows mostly single track.

As well as these variable rail types, the layer also contains information on a variable status of the rail lines, i.e. information on the nature of its use: was it functioning or abandoned, or was its functioning doubtful? Was it added to fill gaps in the network? Figure 10.6(b) shows the data available for the western part of the Netherlands. Still, if one compares the contents of the railroad layer with the existing rail network in the Netherlands it is obvious that some gaps remain. Some can be explained because they refer to bridges and tunnels, which can be found in other layers of the DCW (see Rotterdam). However, some real errors remain; for instance, gaps in the line between Utrecht and Leiden, and in the Rotterdam harbour area. The opposite also occurs. A rail line in the north of the country that was abandoned at the beginning of the 1960s still appears (Figures 10.6e and f). This is due to an error on the original paper maps, which still show the line (Figure 10.6c and d). Most peculiar, however, is the line at Hook of Holland, west of Rotterdam, which goes halfway to England into the North Sea. This error was caused because of unfamiliarity with the area during digitizing. The symbol that represents a railroad is similar to that of the harbour pier at the Hook, which was mistaken for a railroad by the digitizer operator.

From this experiment with DCW data in a well-mapped area it can be concluded that, though metadata is available, one should be careful in using the data. Visualizing the data as well as the meta-information can help the user decide if it is fit for use, and how much work has to be done to edit the data. It should also be realized for what purpose the original

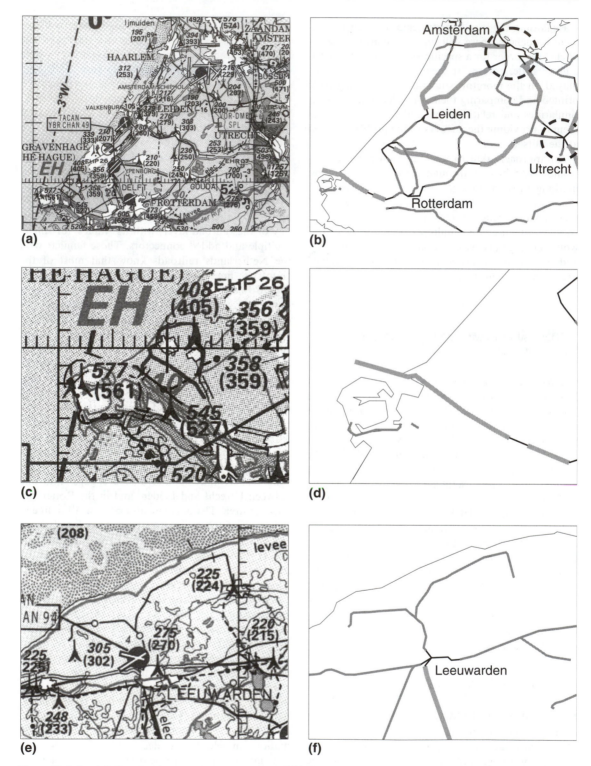

Figure 10.6 Dutch Railroads: (a) western Netherlands from ONC sheet E-2 1985; (b) the same area from DCW's railroad layer; (c) the map around Rotterdam harbour; (d) non-existing railroads in the DCW's railroad layer; (e) northern Netherlands abandoned railroads on the map; and (f) in the DCW railroad layer

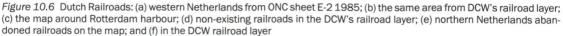

map was compiled (navigation charts for fighter pilots).

10.3.2 East African Highlands

For some applications the DCW data are not available. Surface analysis is such an example. Relief data for the earth's more remote areas could not be retrieved from the original aeronautical charts. However, as can be seen in Figure 10.7(a), the paper maps do provide a shaded relief image of the area, but an overprint warns the user that the heights are believed not to exceed a certain number of feet. The paper maps are probably extrapolated or based on less accurate data. In the DCW database these areas come without contour lines or height points, as is shown by Figure 10.7(b).

10.3.3 Land use and vegetation in Yellowstone National Park

Figure 10.8 shows the contents of the land cover layer (patches) and the vegetation cover layer (lines) at the edge of Yellowstone National Park in the United States. The first contains features relevant for navigation, while the second contains a vegetation cover derived from USGS classified Advanced Very High Resolution Radiometer imagery. As can be seen in Figure 10.8, the image has a 'blocky' appearance because it is derived from raster-based remote sensing data. In some places the data have been smoothed, in others this was not possible because the positional error would become significant. If a user intends to use this type of data in an overlay operation the nature/source of the data should again be considered carefully.

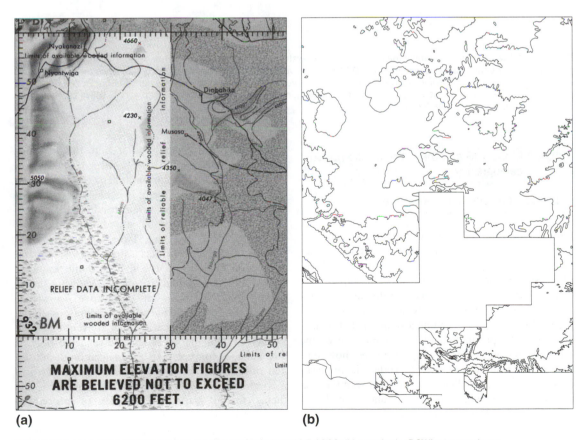

(a)

(b)

Figure 10.7 East African Highlands: (a) detail from ONC sheet M-5 1966; (b) gaps in the DCW's contour layer

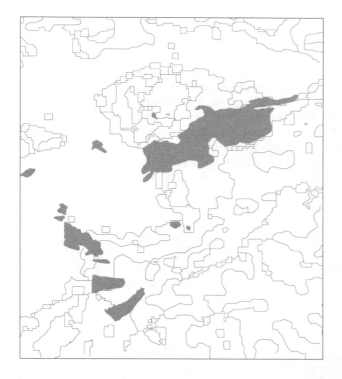

Figure 10.8 The need to know about the data source, Yellowstone National Park and surroundings. The shaded area represents land cover types and is derived from the DCW's land cover layer. It has a vector-based origin, since the original maps were manually digitalized. The lines represent boundaries between vegetation types and were derived from DCW's vegetation cover layer. It has a raster-based origin because the data originate from remote sensing observations, which can be clearly seen just above the largest shaded area

10.4   Maps, GIS and the need for rule-based cartography

10.4.1   Cartographic knowledge

During the 1980s expectations regarding knowledge-based systems ran high. It was thought that these systems could replace an expert in a specific discipline. Similar ideas existed for the cartographic discipline. Especially with the rise of GIS the existence of a cartographic knowledge-based system as part of the GIS software was seen as a solution that would guarantee the production of proper maps, even by users not skilled in cartography. It proved, however, that expectations were too high. Today the application of these knowledge-based systems seems to be restricted to smaller, clearly bound knowledge domains only. The cartographic discipline can be used to illustrate this. First, it proves that a discipline such as cartography is far too large to fit a single knowledge-based system, and second, the discipline is difficult to lay down in a set of rules. If one were to give ten professional cartographers the same data and ask them to map them, it is likely that ten different, but correct, maps would result.

How can knowledge be stored and used in a knowledge-based system? Two knowledge types can be distinguished, declarative knowledge and procedural knowledge. The first is about 'knowing what' and the second is about 'knowing how'. Cartographic examples of declarative knowledge are real objects such as atlases, definitions found in cartographic dictionaries and conventions such as the one that the sea should be rendered blue. It also includes concepts like a classification of atlases, and rules which, for instance, define the choice of a map projection depending on the nature of the theme to be mapped. Examples of procedural knowledge are methods and algorithms, such as the Douglas–Peucker line generalization algorithm. Also, strategies and methodological approaches belong in this catogory. In knowledge-based systems meta-knowledge (knowledge about knowledge, for instance about its quality) is of great importance as well. Declarative knowledge is often stored in a hierarchical structure. The procedural knowledge is stored in production-rules. These rules are of the IF . . . THEN . . . nature. For instance, 'IF this polygon is closed AND has a size less than 0.2, THEN it is a symbol, with a certainty of 0.8'.

A knowlege-based system is structured in three main components. A knowledge base, which stores the declarative and procedural knowledge, a knowledge interpretation engine, which links procedural with declarative knowledge, and a control unit, a user interface to use and update the system. For a more extensive discussion on knowledge-based systems and their role as spatial decision support systems, see Densham (1991). Sometimes, less ambitious knowledge-based systems are called rule-based systems. These systems operate according to a set of fixed rules while a knowledge-based system has the capacity to learn by adding new rules. As well as knowledge- and rule-based systems, cartographers also experiment with case-based reasoning. The application of case-based reasoning in generalization means that cases similar to the current problem are retrieved from a special database, the best match is selected, and if necessary adapted. The generalization process can be executed accordingly.

10.4.2 Applications, problems and potential

It is difficult to feed knowledge-based systems with cartographic knowledge, as is symbolized in Figure 10.9. A cartographer, like many other experts, has some difficulty in wording his/her knowledge. The number of rules needed to describe a specific action is underestimated as well. An experiment at the National Ocean Service's Office of Charting in the United States revealed that, just to make sure a symbol representing a shipwreck would be placed correctly, needed over 200 rules (Bossler *et al.*, 1988).

In many fields of cartography experimental knowledge- or rule-based systems are being developed. Examples found are related to map design, symbol design, name placement, and map revision. Many experiments are also related to generalization. Müller *et al.* (1995) have reported on several experiments. In several of the experiments quoted above it is concluded that organizations with clear mapping tasks, such as topographic mapping organizations, will be able to increase productivity and uniformity throughout their product range through the use of knowledge systems. What about GIS users?

Often GIS users are not interested in cartographic design rules. There are several options available to assist these users anyhow. Because most users will follow defaults offered by the software, the most simple solution is to offer these defaults according to cartographic rules. Examples of default settings for choropleth mapping are the number of classes, the classification method, and the colour range. It should still be possible to divert from these rules in special situations. Another option would be, as can be seen with many of today's Microsoft-Windows software packages, to offer so-called wizard functions. Each stage of the map design process is accompanied by a set of suggestions to help solve potential problems. The user has to click on an icon to access the suggestions. A variation would be that in particular situations these suggestions are forced upon the user, for instance, when the user decides to go for a colour scale instead of a grey value scale for a choropleth map. The user has to consciously make this error by removing the suggestion from the screen. It is also possible to include some map design knowledge in the generic help functions of the software. Here it will only be accessed when the user has asked for help. The option to guide the user along strict lines, without the possibility of straying off a predetermined path, seems less favourable. In an exploratory GIS environment the wizard function approach seems to be the most appropriate.

10.5 Cartography, GIS and spatial information policy

Geographical information, just as geodetic data, does not just happen. It has to be collected, usually at heavy cost to the taxpayer. Sometimes it will be made available to the taxpayers again, at a price. Society decides which data categories will be needed for government, and should therefore be collected. Governments needed topographical maps for defence, census maps for implementing socio-economic policies, resource mapping for physical planning, highway mapping for maintaining that crucial infrastructure, and environmental mapping for offsetting the effects of physical planning and highway building.

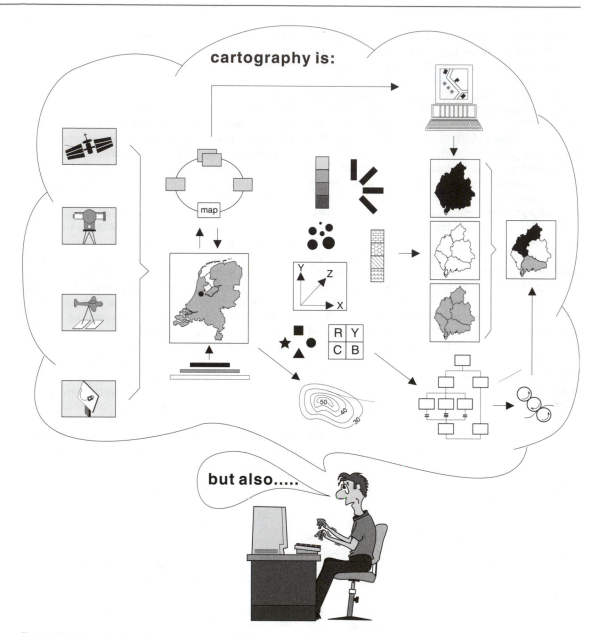

Figure 10.9 The difficulty of feeding a cartographic knowledge base

Nowadays, it is considered wasteful that all the government agencies that engage in collecting spatial data, processing them, putting them in databases and producing maps from these, do so only on their own behalf and do not consult each other. Therefore the trends towards database exchange could be offset by uncontrolled non-standardized construction of databases. As this would force society to pay enormous extra sums because of duplicating existing work, the strategy has been chosen in a number of European and English-speaking countries to formulate spatial information policies. Implementing these spatial information policies would allow for the integration of different databases, developed and kept up to date by different government agencies, which could all be used for solving specific problems (Figure 10.10). By

ACTIVITIES   **ACTORS**

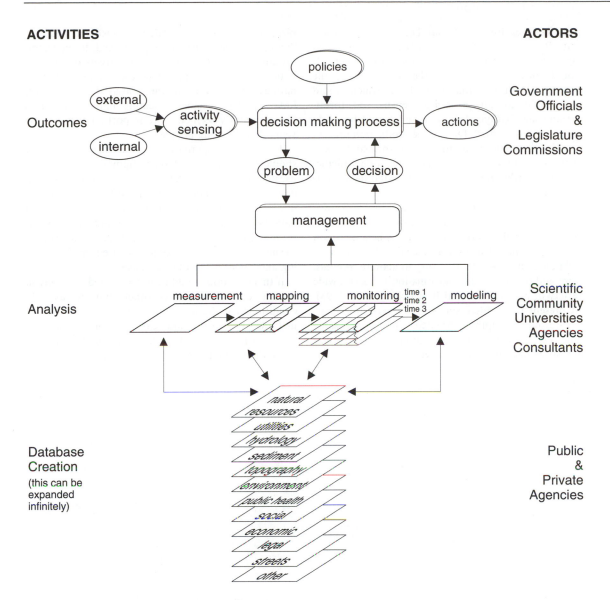

Figure 10.10 Spatial data infrastructure in the United States (from: Spatial data needs: the future of the national mapping program, Washington: National Research Council, 1990)

preparing maps only from databases kept up to date, a number of the problems associated with paper maps could be offset: the maps would not age, show only a limited amount of detail, would not bear street names when not required, or would not be intersected by a map frame.

In the United States the National Mapping Division of the United States Geological Survey (USGS) has been asked to act as administrator of the national spatial data infrastructure. As such, its task would be to produce the 1:24 000 coverage of the coterminous United States in digital format. This is supposed to be completed around the year 2000. Because digital information can be exchanged easily when structured according to a common standard, it is possible that USGS will allow its database to be supplemented with topographic information collected by other parties, albeit checked by USGS officers. The resulting National Digital Cartographic Data Base (see also Section 5.6.4) would contain all topographic object categories in separate layers. It would form the basis for the next project, the National Digital Spatial Data Base, which would additionally contain geological, census, soil, forest, wetland and bathymetric data, and would be ready around the year 2010. During the production process the USGS would have to be transformed from a mapping organization into an organization structuring and coordinating the national spatial data infrastructure (NSDI).

This NSDI is defined as 'the means to arrange geographic information that describes the arrangement and attributes of features and phenomena on the Earth. The infrastructure includes the materials, technology and people necessary to acquire, process, store and distribute such information to meet a wide variety of needs' (National Research Council, 1993). At present the United States is still very much natural-resources-mapping oriented, while at the same time the majority of the population works in service industries and lives in urbanized areas. These urba-

nized areas are not adequately mapped yet, and providing for these urban maps, linked to cadastral inventories, would be one of the targets of the new national digital spatial database. Neither does a nationwide street centreline spatial database exist, linked to street names; locating natural resources has apparently seemed more important through the years than locating people.

The national spatial data infrastructure aimed at should be flexible, available nationwide, and easy to use and to base other activities on. A national spatial data transfer standard should see to it that spatial information can be exchanged easily. Adhering to national data quality standards should ease the collecting of this information by other departments next to the USGS as well, in order to speed up the building and updating of this database.

If this situation would be implemented, not only in the United States, but for other countries as well, conditions for the use of maps as tools for analysis in GISs and for communication would be boosted even more, as it would entail continually available up-to-date spatial information.

References

Anon. (1982) *Atlas van de Nationale Survey*, Vol. II. Brussel: Bestuur van de Stedebouw en Ruimtelijke Ordening, Ministerie van Openbare Werken

Anon. (1991) CD-ROM *Atlas de France*, GIP/RECLUS Montpellier.

Anon. (1992) *Sim City* Maxis.

Anon. (1992) *Times Atlas of the World*. London: Times Books.

Anon, (1994) *Times World Map and Database* (1994). London: Times Books.

Anon. (1995) *GIS sourcebook 1995*. Fort Collins: GIS World (yearly issue).

Anson, R.W. and F.J. Ormeling (1993–1996) *Basic cartography*, Vols 1–3 London: Elsevier.

Bernard, M. and P. Miellet (eds) (1993) Desktop mapping *GIS Europe*, **2**(10), 24–46.

Berry, B.J.L. and D.F. Marble (1968) *Spatial analysis*. Englewood Cliffs: Prentice Hall.

Bertin, J. (1983) *Semiology of graphics*. Madison, WI: University of Wisconsin Press.

Bill, R. (1994) *Multimedia GIS – definition, requirements and applications*. The 1994 European GIS Yearbook. London: Taylor and Francis, pp. 151–154.

Blackford, R., and D. Rhind (1988) The ideal mapping package. In D. Rhind and D.R.F. Taylor (eds), *Cartography, past present and future*. London: Elsevier/ICA, pp. 157–168.

Board, Chr. (1990) Report of the working group on cartographic definitions. *Cartographic Journal*, **29**, 65–69.

Boardman, D. (1983) *Graphicacy and Geography Teaching*. London: Croom Helm.

Bossler, J.D., D.L. Pendleton, G.F. Swetnam, R.L. Vitalo, C.R. Schwartz, S. Alper and H.P. Dunley (1988) Knowledge-based cartography: the NOS experience. *The American Cartographer*, **15**(2), 149–162.

Brassel, K.E. and R. Weibel (1988) A review and cenceptual framework of automated map generalization. *International Journal of GIS*, **2**, 229–244.

Bregt, A.K. (1991) Mapping uncertainty in spatial data. *Proceedings of the Second European Conference on Geographical Information Systems*, Brussels, **1**, pp. 149–154.

Brewer, C.A. (1994) Color use guidelines for mapping and visualization. In A.M. MacEachran and D.R.F. Taylor (eds), *Visualization in modern cartography*, Oxford/New York: Pergamon, pp. 123–147.

Brodlie, K.A. (1994) A typology for scientific visualization. In H.M. Hearnshaw and D.J. Unwin (eds), *Visualization in geographical information systems*. London, Wiley, pp. 34–47.

Brown, A. (1982) A new ITC colour chart based on the Ostwald colour system. *ITC Journal*, **1982**, 109–118.

Brown, A. and P.W.M. Schokker (1989) Offset-printed colour charts for use with a Macintosh II Microcomputer. *ITC Journal*, **1989**(3/4), 225–228.

Burrough, P.A. (1986) *Principle of Geographical Information Systems for land resources assessment*. Oxford: Oxford University Press.

Buttenfield, B.P. and R.B. McMaster (eds) (1991) *Map generalization. Making decisions for knowledge representation*. London: Longman.

Campbell, C.S. and S.L. Egbert (1990) Animated cartography/Thirty years of scratching the surface. *Cartographica*, **27**(2), 24–46.

Chrisman, N.R. (1984) The role of quality information in the long-term functioning of a geographical information system. *Cartographica*, **21**, 79–87.

Clockwork Software (1993) *Millennium 1.21*. Chicago: Clockwork Software.

Cole, J.P. and King, C.A.M. (1968) *Quantitative geography: techniques and theories in geography*. London: Wiley.

Compton's New Media Inc. (1995) *Compton's Interactive Encyclopedia for Windows*. Compton's New Media Inc.

Cowen, D.J. (1988) GIS versus CAD versus DBMS: what are the differences? *Photogrammetric Engineering and Remote Sensing*, **54**(11), 1441–1555.

Dahlberg, R.E. (1967) Towards the improvement of the dot map. *International Yearbook of Cartography 1967*, pp. 157–167.

Dale, P.F. (1991) Land information systems. In D.J. Maguire, M.F. Goodchild and D. Rhind (eds) *Geographical Information Systems*. London: Longman.

Densham, P.J. (1991) Spatial decision supports systems. In D.J. Maguire, M.F. Goodchild and D. Rhind (eds) *Geographical information systems*. London: Longman, pp. 403–412.

Dent, B.D. (1985) *Principles of thematic map design*. Reading MA: Addison Wesley.

Depuydt, F. (1989) *Elkab IV Topographie. Archaeological–topographical surveying of Elkab and surroundings*. Uitgaven van het Comité voor Belgische opgravingen in Egypte, Brussels.

DiBiase, D. (1990) Visualization in earth sciences. *Earth and Mineral Sciences, Bulletin of the College of Earth and Mineral Sciences, The Pennsylvania State University*, **59**(2), 13–18.

DiBiase, D., A.M. MacEachren, J.B. Krygier and C. Reeves (1992) Animation and the role of map design in scientific visualization. *Cartography and GIS*, **19**(4), 201–214.

Digital Cartographic Data Standard Task Force (DCDSTF) (1988) The proposed standard for digital cartographic data. *The American Cartographer*, **15**(1).

Douglas, D.M. and T.K. Peucker (1973) Algorithms for the reduction of the number of points required to represent a digitized line or its caricature. *Canadian Cartographer*, **10**(3), 112–122.

Electromap Inc. (1991) *InfoNation 2.0*. Fayetteville, AR: Electromap.

ESRI (1993) *Digital Chart of the World for use with Arc/Info software*. Redlands: ESRI.

Fischer, P. (1994) Randomization and sound for the visualization of uncertain spatial information. In D. Unwin and H. Hearnshaw (eds) *Visualization in Geographic Information Systems*. London: Wiley, pp. 181–185.

Freitag, U. (1992) Cartographic conceptions: contributions to theoretical and practical cartography 1961–1991. *Berliner geowissenschaftliche Abhandlungen Reihe C Bd 13*.

Gächter, E. (1969) Die Weltindustrieproduktion 1964. Eine statistisch-kartographische Untersuchung des sekundären Sektors, Zürich.

Geels, J-H. (1987) Eeen model voor de keuze van vlaksymbolen. *Kartografisch Tijdschrift*, **13**(4), 22–27; **14**(1), 18–23.

Grünreich, D. (1990) Konzeption und erste Erfahrungen aus der Ausbauphase des digitalen Landschaftsmodell 1:25 000 (DLM25). Proceedings XIX FIG Congress, Helsinki, **5**, pp. 152–162.

Goodchild, M. and S. Gopal (eds) (1989) *Accuracy of spatial databases*. London: Taylor & Francis.

Guptill, S.C. and L.E. Starr (1984) *The Future of Cartography in the Information Age*. Washington: ICA.

Healey, R.G. (1991) Database management systems. In D.J. Maguire, M.F. Goodchild and D. Rhind (eds) *Geographical Information Systems*. London; Longman, pp. 251–267.

Hearnshaw, H.M. and D.J. Unwin (1994) *Visualization in geographic information systems*. London: Wiley.

HMSO (1980) *People in Britain, a census atlas*. London: Census Research Unit, HMSO.

Hootsmans, R.M. and F.J.M. Van Wel (1992) Kwaliteitsinformatie ter ondersteuning van de integratie van ruimtelijke gegevens. *Kartografisch Tijdschrift*, **18**(2), 49–55.

Hootsmans, R.M. and F.J.M. Van der Wel (1993) Detection and visualization of ambiguity and fuzziness in composite spatial datasets. *Proceedings of the Fourth European Conference on Geographical Information Systems*, **2**, Utrecht, pp. 1035–1046.

Imhof, E. (1972) *Thematische kartographic*. Berlin: De Gruyter.

Imhof, E. (1982) *cartographic relief representation (English translation)*. New York: W. de Gruyter. First published in German in 1965.

Jenks, G.F. and M.R.C. Coulson (1963) Class intervals for statistical maps. *International Yearbook of Cartography 1963*, pp. 119–133.

Kadmon, N. (1992) *An introduction to toponymy. Theory and practice of geographical names*. Pretoria: University of Pretoria Department of Geography.

Keates, J.S. (1989) *Cartographic design and production*, 2nd edition. Harlow: Longman.

Koussoulakou, A. (1990) *Computer-assisted cartography for monitoring spatio-temporal aspects of urban air pollution*. Delft: Delft University Press.

Kraak, M.J. (1988) *Computer-assisted cartographical three-dimensional imaging techniques*. Delft: Delft University Press.

Kraak, M-J. (1994) Interactive modelling environment for three-dimensional maps: functionality and interface issues. In A.M. MacEachren and D.R.F. Taylor (eds), *Visualization in modern cartography*. Oxford: Pergamon, pp. 269–286.

Kraak, M-J. and A.M. MacEachren (1994) Visualization of the temporal component of spatial data. In T.C. Waugh (ed.), *Advances in GIS research. Proceedings 6th International Symposium on Spatial Data Handling*. London: Taylor and Francis, pp. 391–409.

Kraak, M.J., J.C. Müller and F.J. Ormeling (1995) GIS cartography: visual decision support for spatio-temporal data handling. *International Journal of Geographic Information Systems*, **9**,6, pp. 637–645.

Krygier, J. (1994) Sound and cartographic visualization. In A.M. MacEachran and D.R.F. Taylor (1994) *Visualization in modern cartography*. Modern Cartography series. Oxford: Pergamon.

Laurini, R. and D. Thompson (1992) *Fundamentals of spatial information systems,* The APIC Series no. 37. London: Academic Press.

MacDougall, E.B. (1992) Exploratory analysis, dynamic statistical visualization, and geographic information systems. *Cartography and GIS,* **19**(4), 237–246.

MacEachren, A.M. (1994) Visualization in modern cartography: setting the agenda. In D.R.F. Taylor and A.M. MacEachren (eds) *Visualization in modern cartography.* London: Pergamon Press.

MacEachren, A.M. and D.R.F. Taylor (1994) *Visualization in modern cartography.* Modern cartography series. Oxford: Pergamon.

Maguire, D.J., M.F. Goodchild and D. Rhind (1991) *Geographical Information Systems.* London: Longman.

Makower, J. (1992) *The map catalog.* New York: Vintage Books.

Marx, R.W. (1990) The TIGER system: automating the geographic structure of the United States. In D.J. Peuquet and D.F. Marble (eds) *Introductory readings in geographic information systems.* London: Taylor and Francis.

McCormick, B., T. DeFanti and M. Brown (1987) Visualization in scientific computing. *Computer Graphics,* **21**, 6.

McMaster, R.B. and K.S. Shea (1988) Cartographic generalization in a digital environment: a framework for implementing in a GIS. *Proceedings GIS/LIS'88,* San Antonio, pp. 240–249.

McMaster, R.B. and K.S. Shea (1992) *Generalization in digital cartography.* Washington, DC: Assocn of American Geographers.

Moellering, H. (1983) Designing interactive cartographic systems using the concepts of real and virtual maps. *Proceedings Autocarto 6,* **2**, 53–64.

Monkhouse, F.J. and H.R. Wilkinson (1971) *Maps and diagrams.* London: Methuen.

Monmonier, M. (1992) Time and motion as strategic variables in the analysis and communication of correlation. In P. Bresnahan et al. (eds) *Proceedings 5th International Symposium on Spatial Data Handling.* Charleston, South Carolina, pp. 72–81.

Morrison, J.L. (1974) A theoretic framework for cartographic generalization with emphasis on the process of symbolization. *International Yearbook of Cartography,* **14**, 115–127.

Müller, J-C. (1991) Rule-based generalisation: potential and impediments. *Proceedings 4th International Symposium on Spatial Data Handling,* Zürich, pp. 557–571.

Müller, J.C., J.P. Lagrange and R. Weibel (eds) (1995) *GIS and generalization, methodology and practice.* London: Taylor and Francis/European Science Foundation.

National Research Council (1990) *Spatial data needs: the future of the National Mapping Program.* Washington, DC: National Academy Press.

National Research Council (1993) *Towards a coordinated spatial data infrastructure.* Washington, DC: National Academy Press.

Neurath, O. (*c.* 1930) *Atlas Gesellschaft und Wirtschaft.* Leipzig: Bibliographisches Institut.

Ormeling, F.J., B.J. Kobben and R. Perez Gomez (1996) *Proceedings of the Seminar on Teaching Animated Cartography, Madrid 1995.* Utrecht: International Cartographic Association.

Peuquet, D.J. (1984) A conceptual framework and comparison of spatial data models. *Cartographica,* **21**(4), 66–113.

Pillewizer, W. and T. Töpfer (1964) Das Auswahlgesetz, ein Mittel zur kartographischen Generalisierung. *Kartographische Nachrichten,* **14**(4).

Raper, J. (ed.) (1989) *Three-dimensional applications in GIS.* London: Taylor and Francis.

Rhind, D. (1992) The information infrastructure of GIS. In P. Bresnahan *et al.* (eds), *Proceedings 5th International Symposium on Spatial Data Handling,* Charleston, pp. 1–19.

Robinson, A.H. (1974) A new map projection: its development and characteristics. *International Yearbook of Cartography,* **14**, 145–155.

Robinson, A.H. and R.E. Bryson (1957) A method for describing quantitatively the correspondence of geographical distributions. *Annals of the Association of American Geographers,* **47**(4).

Robinson, A.H., J.L. Morrison, P.C. Muhrcke, A.J. Kimerling and S.C. Guptill (1995) *Elements of cartography,* 6th edition. New York: Wiley.

Roelfsema, C.M., A. van Voorden and F. van Tatenhove (1995) Vormverandering van gletsjers. *Geodesia,* **37**(4), 193–198.

Samet, H. (1990) *The design and analysis of spatial data structures.* Reading, MA: Addison-Wesley.

Sensus Pertanian (1973) Agricultural Census, Biro Pusat Statistik, Central Statistical Office, Jakarta 1976.

Siekierska, E.M. (1993) From the electronic atlas system to the electronic atlas products: Electronic atlas of Canada from the beginning to the end. In I. Klinghammer *et al.* (eds), *Proceedings of the seminar on Electronic Atlases held in Visegrad 1993.* Budapest: Eötvös Lorand University, pp. 103–111.

Smith, R.M. (1989) *Atlas of Arkansas.* Fayetteville: University of Arkansas Press.

Smith, R.M. and T. Parker (1995) An electronic atlas authoring system. *Cartographic Perspectives,* **20**, 35–39.

Snyder, J.P. (1987) *Map projections–a working manual.* US Geological Survey Professional Paper 135.

Software Toolworks (1994) *World Atlas.* Novato, CA: Software Toolworks.

Spiess, E. (1993) *Schweizer Weltatlas.* Zürich: Lehrmittelverlag Kanton Zürich.

Taylor, D.R.F. (1991) Geographic information systems: the microcomputer and modern cartography. In D.R.F.

Taylor (ed.) *Geographic Information Systems*. Oxford/New York: Pergamon.

Taylor, D.R.F. (1994) Cartographic visualization and spatial data handling. In T.C. Waugh (ed.), *Advances in GIS research*, Proceedings 6th International Symposium on Spatial Data Handling. London: Taylor and Francis, pp. 16–28.

Tobler, W.R. (1973) Choropleth maps without class intervals? *Geographical Analysis*, **1973**(5), 262–265.

Tomlin, C.D. (1990) *Geographic information systems and cartographic modelling*. Englewood Cliffs: Prentice Hall.

Van Elzakker, C.P.J.M. (1993) The use of electronic atlases. In I. Klinghammer *et al.* (eds), *Proceedings of the seminar on Electronic Atlases held in Visegrad 1993*. Budapest: Eötvös Lorand University, pp. 145–157.

Van der Wel, F.J.M. and R.M. Hootsmans (1993) Visualization of quality information. In P. Mesenburg (ed.) *Proceedings 16th International Cartographic Conference*, Cologne 1993. Bielefeld: Deutsche Gesellschaft für Kartographie.

Wood, M. (1994) Visualization in a historical context. In MacEachren, A.M. and D.R.F. Taylor (1994) *Visualization in modern cartography*. Modern Cartography series. Oxford: Pergamon, pp. 13–26.

INDEX

Note: Illustration page references appear in italic